Discovery and Decision

Discovery and Decision

Exploring
the Metaphysics and Epistemology
of Scientific Classification

Rebecca Bryant

Madison • Teaneck
Fairleigh Dickinson University Press
London: Associated University Presses

© 2000 by Associated University Presses, Inc.

All rights reserved. Authorization to photocopy items for internal or personal use, or the internal or personal use of specific clients, is granted by the copyright owner, provided that a base fee of $10.00, plus eight cents per page, per copy is paid directly to the Copyright Clearance Center, 222 Rosewood Drive, Danvers, Massachusetts 01923. [0-8386-3876-7/00 $10.00 + 8¢ pp, pc.]

Associated University Presses
440 Forsgate Drive
Cranbury, NJ 08512

Associated University Presses
16 Barter Street
London WC1A 2AH, England

Associated University Presses
P.O. Box 338, Port Credit
Mississauga, Ontario
Canada L5G 4L8

The paper used in this publication meets the requirements of the American National Standard for Permanence of Paper for Printed Library Materials Z39.48-1984.

Library of Congress Cataloging-in-Publication Data

Bryant, Rebecca, 1970–
   Discovery and decision : exploring the metaphysics and epistemology of scientific classification / Rebecca Bryant.
     p. cm.
   Includes bibliographical references (p.) and index.
   ISBN 0-8386-3876-7 (alk. paper)
   1. Classification of sciences. 2. Metaphysics. 3. Knowledge, Theory of. I. Title
BD241.B785 2000
001'.01'2—dc21   00-037169

PRINTED IN THE UNITED STATES OF AMERICA

For Steffan

It is now evident that where one discipline ends and the other begins no longer matters, for it is in the nature of the case that the boundaries are ill-defined.
—Patricia Churchland, *Neurophilosophy*, p. x.

We find one truth and embrace it. Then we close our eyes to everything else. It avoids confusion.
—David Eddings, *The Elenium*, p. 164.

Contents

Acknowledgments	9
Glossary	11
1. Introduction: Complexity and the Natural World	15
2. Objectivism	20
3. Internal Realism I	29
4. Internal Realism II: Criticisms and Implications	38
5. The Psychology of Categorization	51
6. Philosophy and the Psychology of Categorization	66
7. Five Interrelated Theses: The Theory Explained	73
8. Philosophical Contexts I: Critiques of Natural Kinds	88
9. Philosophical Contexts II: Contemporary Philosophy of Biology	98
10. Concluding Remarks: Limitation and Classification	111
Notes	117
Bibliography	125
Index	129

Acknowledgments

The original research on which this book is based was supported by a Vans Dunlop Scholarship from the University of Edinburgh's Faculty of Arts Scholarship Fund, for which I am very grateful.

My warm thanks also go to Peter Lewis, Alexander Bird, John Dupré and David Bloor, all of whom guided me in improving much earlier versions of the text.

Glossary

Classification or *categorization:* The process by which human beings group together particular entities and treat them as equivalent in some sense or senses. We group all birds together, for example, treating them as equivalent because they all have wings, feathers and beaks.

Class, category or *kind:* A number of entities which have been grouped together by the process of classification or categorization. The collection of animals, all of which have wings, feathers and beaks, is known as the 'bird' class or category.

Concept: A mental representation or mental counterpart of a class, category or kind. The mental representation or idea that human beings have of birds will correspond to the 'bird' class or category.

Discovery and Decision

1
Introduction:
Complexity and the Natural World

In early 1999, a debate erupted—a debate involving astronomers and their beliefs about how Pluto should be classified. On its discovery in 1930, Pluto was designated a planet, but recently its planetary status has been questioned. The argument goes thus: Pluto is so small and so distant in its orbit around the Sun that it does not count as a proper planet. Instead, it should be viewed as the largest member of a group of small objects—known as Trans Neptunian Objects (TNOs)—which orbit at great distance around the Sun.

But opinions remain divided. Astronomer Nevile Kidger argues that the reasons for downgrading Pluto's status are compelling—its mass is only about one fifth that of the Moon and so it is very difficult to claim that Pluto is really a planet. Other astronomers argue that Pluto is more massive by about a factor of 100 than any known TNO, with the exception of its satellite Charon—and Pluto and Charon together make up more than fifty percent of the mass of all TNOs discovered to date. In this case, does it really make sense to say that Pluto is nothing more than a TNO? Others opt for the middle path. Gareth Williams of the Minor Planet Center in Massachusetts (which catalogs the orbits of solar system objects) has suggested not demotion, but 'dual citizenship' for Pluto.

A quandary has arisen. Is Pluto a planet? Is it a TNO? Or could it, just possibly, be both? It is unclear. The natural world does not provide an answer. On the basis of Pluto's properties, we can evince arguments in favor of all three positions.

As it turns out, the International Astronomical Union eventually decided not to demote Pluto from its planetary status, claiming that dual citizenship for the planet would have diminished its standing. But this kind of example draws our attention to a feature of science we do not usually see—the natural world is not as simple or clear-cut as we tend—or are led—to believe and, as a result, disputes arise among scientists concerning the proper description or classification of natural entities like Pluto.

Common sense views about the straightforwardness of nature and science's role as one of mirroring nature's inherent properties conform to a philosophical tradition dating back (at least) to Aristotle. This tradition claims that the natural world possesses a unique structure in and of itself. More specifically, it emphasises that the world divides inherently into classes and that members of each class share a common essence or nature which takes the form of a set of defining characteristics. The essence comprises *what it is* to be a member of the relevant class or kind. Astronomers might argue, for example, that in order to qualify as a member of the class of planets, an object must be spherical, above a certain size and traverse an orbit relatively close to the Sun. This doctrine of essentialism of kinds has implications for both classification and definition. It is argued that to classify entities correctly is simply to reproduce the world's inherent structure. To give the correct definition of what it is to be a particular kind of thing and so to provide the meaning of the word which refers to that kind of thing is to reiterate the essence or the defining characteristics of the natural class to which it belongs.

The doctrine of essentialism entails a number of cognate assumptions. These include the notions that correct classification is a matter of metaphysics alone, that epistemology is irrelevant for matters of classification and that human beings play an entirely passive role in the classificatory process by simply *reflecting* the metaphysics.

More recently, this doctrine has been adopted in the early work of Hilary Putnam and by Saul Kripke. These philosophers have linked essentialism with *science,* claiming that it is the role of science to uncover the essential properties which underlie and explain the natural world's inherent division into kinds.

The aim of this book is to challenge this doctrine of essentialism of kinds which is so often taken as characteristic of scientific classification. My starting point is the *complexity* of the natural world and I argue that, as a result of this complexity, numerous scientifically significant patterns of similarity or regularity exist according to which natural entities can be divided into kinds or classes. Such variety can give rise to disputes over classification of particular entities (as in the Pluto case) as well as to competing schemes of classification. It also produces a situation in which scientists are forced to *choose* which among a number of different patterns of similarity or regularity are to count as salient for classification. As such, I present a *pluralist* account of classification—one in which not just one, but several systems for grouping entities into kinds can be correct, true or reflective of (some portion of) the natural world. In particular, my aims are:

- To illustrate that *scientific* classification comprises a *synthesis of metaphysics and epistemology.*
- And, in view of this synthesis, to show that the *psychological explanation-based account of categorization is a suitable model of expert scientific as well as of lay classification.*

I concentrate on scientific classification partly because of the weight which Putnam and Kripke lend to science in explicating their doctrines, and partly because it is natural to expect that if the doctrine of essentialism and its cognate assumptions were correct, they would be proved so in the scientific domain, if anywhere, simply because this domain is popularly (and often philosophically) construed as objective and context-free.

Nonetheless, my pluralist account *is not* correctly described as relativist, subjectivist or anti-realist. It represents instead a middle road between the extremes of metaphysical realism and relativism. I do not argue that there is no external world nor that we humans somehow construct the world in which we reside. My position very much demands the existence of a mind-independent world populated by real entities and real properties. It is correctly described as a realist position.

The next three chapters provide a theoretical backdrop against which my account of scientific classification comes to life. Chapter 2 discusses Objectivism, a philosophical doctrine concerning both the way the world is and our understanding of that world, a doctrine which motivates traditional accounts of categorization and essentialism about kinds. Chapters 3 and 4 examine Putnam's internal realism, which opposes Objectivism, yet avoids the extremes of relativism by insisting on the reality and independence of the external world. In chapter 3, I suggest that internal realism's theory of truth fails, arguing instead that we need to stick to the notion of truth as correspondence. Chapter 4 defuses some recurring objections to internal realism and then moves on to explain both how this doctrine accommodates my account of scientific classification and how my account relates to notions of objectivity and realism propounded by philosophers of science Helen Longino and Ronald Giere.

Chapters 5 and 6 discuss research in the psychology of categorization, examining the implications of this work for philosophy. I argue, in particular, that progression from earlier to later psychological positions marks a shift from the interpretation of categorization as a tightly defined process in which humans play an entirely passive role to that of a much more flexible, context-sensitive process in which humans play an instrumental role. Chapter 6 considers the criticisms which an objectivist philosopher—Georges Rey—levies against the psychology of categorization and discusses responses to these criticisms, especially in relation to the division which Rey draws between metaphysics and epistemology. We begin to see how the explanation-based model of categorization—in which metaphysics and epistemology merge—might function as a model of expert scientific classification.

The remaining chapters are concerned with presenting my pluralist account of scientific classification. Chapter 7 makes five interrelated claims about scientific classification—it is not clear-cut, it involves more than the way the world is, it involves human input, it comprises a synthesis of metaphysics and epistemology, and the explanation-based model applies to expert scientific classification—illustrating each claim by reference to three socio-historical case studies which catalog disputes in scientific taxonomy. Chapters 8 and 9 relate my account of scientific classification to similar positions held by other philosophers. Chapter 8 focuses on critiques of the Putnam/Kripke approach to natural kinds from Keith Donnellan and John Canfield, while chapter 9 examines the pluralist understanding of natural kinds shared by John Dupré, Philip Kitcher and Ian Hacking. And chapter 10 confirms the realism entailed by my pluralist account, discussing two important constraints on scientific classification.

Finally, a note about methodology is perhaps in order. This book—unusually for a work in philosophy—comprises an interdisciplinary, empirical study, which embraces both psychological evidence relating to the process of categorization and empirical material (historical and contemporary) concerning discovery and theory in scientific classification. What are the reasons for this? First, I consider that, if we want to find out how things are in the world, we must heed what empirical, scientific studies tell us. And scientific taxonomy—both past and present—points to the fact that the world is simply too complex to admit of unique classifications of the entities within it. If we accept that science provides the best available description of the natural world, we must also accept that what science tells us will not necessarily reflect *a priori* conceptions about that world, such as those encapsulated in the doctrine of Objectivism. Second, I concur with Ronald Giere (among others) that philosophers of science rest on a par with scholars in psychology and sociology who study science itself *scientifically*.[1] On this model, the philosopher of science becomes a theoretician of the science of science—and I consider myself a philosopher of science in this tradition. In claiming that the explanation-based model applies to expert scientific classification, I am propounding a theory about cognitive processes underlying the activity of scientific classification—a theory that either stands or falls in the light of evidence concerning how scientists engage in classification. I believe it stands in the light of the relevant evidence. My goal, then, is to further understanding of the way in which scientists do science, not to inform scientists of how, ideally, they ought to do their job.

2

Objectivism

Differing theoretical backdrops motivate differing accounts of classification. Objectivism—a theory which speaks about both the way the world is (metaphysics) and our experience of that world (epistemology)—underpins the essentialist or classical view of classification. Before presenting any alternative to the classical view, we need to explore the objectivist backdrop, to understand its aims, to pinpoint its proponents, and to highlight its shortcomings.

The most basic objectivist tenet involves a commitment to the existence of a real world, a world which is independent of and external to human minds and their experience. Objectivists assume that the independent world consists of entities plus properties and relations holding between those entities. They also assume that the world is *uniquely* structured in terms of these entities, properties and relations. According to the objectivist worldview, there is *one and only one correct description of reality*—the one that mirrors the structure inherent in the world.

Along with the primary metaphysical notion that reality consists of entities and fixed properties and relations, there often goes a further assumption, that some of the properties which these entities exhibit are *essential*. Possession of essential properties is supposedly what makes the entity the thing that it is—the properties capture the essence of the thing. Other properties are accidental and so do not capture the essence of the thing. We might say, for example, that the property of being a rational featherless biped captures the essential nature of human beings. Other properties such as height and skin color are merely accidental. They do not contribute to the essential nature of human beings.

The classical view of categorization states that all entities sharing the same given properties, or set of properties, form a group or category. These common properties in fact define the category and provide necessary and sufficient conditions for membership, with each essential property as singly necessary, and the totality of properties as jointly sufficient, for membership in the category. Being rational, featherless and bipedal might thus form the necessary and sufficient conditions for membership in the category of human beings.

Since objectivists maintain that entities and their properties objectively exist, and since the classical view of categorization rests on the (non)possession of properties by entities, the further claim is made that the categories themselves exist objectively. And so the categories into which entities fall also contribute to the structure of the objective world.

In some cases, the classical view of categorization and the doctrine of essentialism combine to produce a special type of category known as a natural kind. The basis for membership in a natural kind (a classical category) is possession of particular essential, rather than accidental properties. The essence which attaches to the natural kind is therefore not a matter of the way language works, but a matter of the inherent nature and structure of the world. Natural phenomena such as animals, plants, and chemical elements and compounds are commonly cited as examples of natural kinds.

Objective logical relations between categories also play an important part in the objectivist world picture. Certain categories are included within other categories, with each included category exhibiting the properties of the inclusive category. Thus an objectivist might claim that the category of human beings falls within the category of mammals and, if mammals are defined as 'warm-blooded creatures which suckle their young,' then all human beings must also be warm-blooded and suckle their young. Or, alternatively, two categories may be mutually exclusive, with no overlap of definition and so no common members. 'Mammals'—defined as 'warm-blooded creatures which suckle their young'—and 'reptiles'—defined as 'cold-blooded creatures possessing bodies covered with scales or bony plates'—represent two such mutually exclusive categories.

Objectivist metaphysics also embraces real-world atomism. Certain properties in the world are taken as basic (atomic) and cannot be broken down or analyzed any further—they have no internal structure. These properties can combine logically into more complex properties which do have internal structure. Being human, for example, may be construed as a complex property comprising a combination of the atomic properties of being rational, featherless and bipedal.

The objectivist's world (objectivist metaphysics) is self-contained and inherently structured. Its entities, properties, categories and relations remain fixed. This internal structure is an *objective feature of the world*, a feature quite *independent of human thought, experience or cognition*.

What, then, does objectivist epistemology look like? How does the objectivist answer questions about meaning, valid reasoning or knowledge of the world? The key is reflection. Our minds, according to the objectivist, can mirror the structure inherent in the world.

Words and mental representations (or concepts) are understood by the objectivist as abstract, meaningless symbols. Only by successfully referring to elements of the objective world do these symbols acquire meaning. When we reason correctly, we manipulate these symbols according to logical rules so that they reflect the structure of the world, and we thereby gain knowledge of the world. Symbols must correspond to objective entities and categories and we must be able to reproduce the logical relations existing between these entities and categories. Correct reasoning and knowledge therefore involve *reflecting the inherent and logical structure of the world*.

Objectivism moves rationality and knowledge away from the personal, human sphere. It assumes the plausibility of the universal, context-free God's Eye perspective. And ultimate (objective) knowledge and rationality become quite independent of individual human thought, belief or cognition.

How do human concepts fit into such an account of human cognition? Concepts, for the objectivist, are nothing more than mental representations of the categories and entities of the world. They must, by definition, exclude anything which depends on human conceptualisation, thought or belief about the world. The classical view of categorization thus governs conceptual categories as well as real-world categories. Conceptual categories comprise symbolic representations of categories in the real or other possible worlds and their members comprise symbolic representations of entities in the real or other possible worlds. Conceptual categories mirror the structure of real-world categories, hence they are defined in terms of necessary and sufficient conditions which are based on the (non)possession of properties by the symbolic representations of real-world entities. These properties and conditions will, of course, mirror the properties and conditions which hold in the real or other possible worlds. Conceptual categories simply are mirror images of real-world categories.

The objectivist account of language runs along similar lines. Statements are meaningful only if they can be evaluated true or false. Linguistic expressions acquire meaning only by corresponding to the real or some possible world. This takes the form of correctly referring, in the case of noun phrases, or of being true or false, in the case of sentences. Referring expressions correspond to entities and predicates correspond to properties. Sentences are true when their noun phrases and predicates relate to one another in such a way as to correspond to some state-of-affairs in the actual or some possible world.

Meaning, for the objectivist, can be explained in terms of truth and truth conditions. Giving the meaning of a particular sentence amounts to giving the conditions under which it would be true (its conditions of satisfaction)—the state-of-affairs in the world which would have to obtain in order for that sentence to be satisfied.

Objectivist epistemology, then, insists that 'correct' cognition and language are quite independent of human thought and activity. Knowledge involves nothing more than accurate world reflection and meaning is simply an objective relation between symbols and the world. In order to gain knowledge or to use language meaningfully, we must adopt a God's Eye, rather than a human, perspective.

OBJECTIVISM IN THE FLESH

I have given a formal characterization of Objectivism in the preceding paragraphs, a characterization that incorporates the various facets of both objectivist metaphysics and objectivist epistemology. Of course, no one philosopher adheres to each and every aspect of the doctrine, but different thinkers have embraced different aspects from the time of Aristotle right up to the present day. We can therefore put some flesh on the bones of Objectivism by casting a glance towards some of its more prominent adherents.

Modern versions of essentialism date back to Aristotle's teachings and the psychological classical view of categorization (to be discussed in chapter 5) derives pretty much directly from his work. In the *Categories* it becomes clear that, for Aristotle, definition is a matter of fitting the thing to be defined into a particular classification scheme. This scheme sorts entities into various kinds (the genera) which are themselves sorted into various sub-kinds (the species) which are distinguished from one another by a number of characteristic properties (the differentia). Giving the correct definition of what it is to be a human, for instance, would involve providing the genus and the differentia of the species. We might say that the species 'human' belongs to the genus 'animal' and is differentiated from other species of this genus by the possession of 'rationality'. A human is therefore a rational animal—these properties capture the essence of humanity.

Aristotle seems to believe that this—the only truly scientific method of classification—reflects the natural order of things and, as such, can yield only *one* correct result for a given entity. In *De Partibus Animalium I*, he outlines the way in which the natural world ought to be divided, saying "it is necessary to divide by privation,"[1] "one should divide by what is in the being, and not by the essential accidents,"[2] and "one should divide by opposites."[3] He also insists that "we should if possible say that because this is what it is to be a man, therefore he has these things; for he cannot be without these parts."[4] And "the essence is not produced; for this is that which is made to be in something else by art or by nature or by some capacity."[5] For Aristotle, objectivist metaphysics and objectivist epistemology go hand-in-hand.

John Locke, by contrast, draws a distinction between the real and nominal essence of things, the real essence comprising "the being of anything, whereby it is what it is" and the nominal essence comprising "that abstract idea which the general, or sortal . . . name stands for."[6] He believes that this inner, real essence causes the phenomenal properties of material things. Yet, since real essences are not visible to the naked eye, we are forced to group things together on the basis of their phenomenal similarities. Even if we *were* able to discover real essences, Locke doubts that they would correspond to the nominal essences and kinds which we have set up. He therefore endorses objectivist metaphysics but, due to our human limitations, rejects objectivist epistemology. Had he lived to see the invention of the microscope and subsequent scientific advances, he may, of course, have endorsed objectivist epistemology too.

Ludwig Wittgenstein focuses, in his *Tractatus*, on language as a means of representing how things are in the world. The world is the totality of facts, he tells us. Facts are more and less complex, but in the last instance, there are so-called atomic facts which cannot be further broken down into simpler facts and which are mutually independent. He construes our language as corresponding isomorphically to the facts of the world. Thus propositions are correspondingly more and less complex, but in the last instance there are so-called atomic propositions which correspond to atomic facts. Wittgenstein's famous metaphor of language as a picture depicting reality depends upon the notions of a one-to-one correspondence between the elements of the picture and the thing pictured and on a common structure shared by language and reality. "We picture facts to ourselves," he says. "A picture presents a situation in logical space, the existence and non-existence of states of affairs."[7]

In accord with Aristotle, the early Wittgenstein upholds both objectivist metaphysics and objectivist epistemology, explaining that "if all true elementary propositions are given, the result is a complete description of the world. The world is completely described by giving all elementary propositions, and adding which of them are true and which false."[8] Language works, Wittgenstein believes, because it is anchored in reality by means of logical isomorphism.

More recently, Hilary Putnam and Saul Kripke have both famously advocated an objectivist interpretation of science, claiming that science's role is one of uncovering the essential properties which underlie and explain the world's inherent division into kinds.[9] It is precisely this understanding of science and scientific classification that I want to counter over the coming chapters.

Putnam challenges two components of earlier theories of meaning: (1) the idea that knowing the meaning of a term is simply a matter of being in a certain psychological state, and (2) the notion that a term's meaning determines its extension. He aims to show that the meaning of natural kind terms depends much more on the real nature of things than earlier theories allow.

Two things count as members of the same kind for Putnam only if they possess the same "important physical properties." In order to determine this, we must first determine what language users are referring to when they use a particular term and then we must uncover the nature of the referent of that term. We can determine, for instance, that when speakers of English talk about 'water,' they are referring to a clear, tasteless, odourless liquid which is found in pools, lakes and waterfalls. We (or our scientists) can then discover that the nature or important physical property of that substance is being H_2O. All liquids which we refer to as 'water' must comprise H_2O molecules. If they do not, our use of the term is incorrect. For Putnam, "once we have discovered that water (in the actual world) is H_2O, *nothing counts as a possible world in which water isn't H_2O.*"[10]

Putnam invites us to imagine a place called Twin Earth which is exactly similar to Earth. Twin Earth has rivers, waterfalls and lakes which are full of a liquid which the Twin Earthians call 'water' and which has all the surface characteristics possessed by Earthian water. On closer inspection, however, Twin Earthian water is found not to have the microstructure H_2O, but a different microstructure, XYZ. Twin Earthian water cannot be water, Putnam insists, because it does not exhibit the essential H_2O microstructure—superficial similarity alone will not do.

Along with Putnam's objectivist metaphysics goes an objectivist epistemology. He asks us to imagine the Twin Earthian situation back in 1750, before the microstructure of water was known. He suggests that *even though we would have had no way of telling Earthian and Twin Earthian water apart,* still "the extension of the term 'water' was just as much H_2O on Earth in 1750 as 1950; and the extension of the term 'water' was just as much XYZ on Twin Earth in 1750 as in 1950."[11] Likewise, although our methods of identifying gold are now much more advanced than they were in ancient Greece, still the extension of 'χρυσὸς' in ancient Greek is the same as the extension of 'gold' in modern English. Even though the Greeks might have called various pieces of metal (which did not have atomic number 79) 'χρυσὸς', they would have been mistaken in doing so. What comprises the essence of a natural kind is, for Putnam, quite independent of human knowledge, beliefs and interests—it is rather a matter of the way reality is.

Kripke, like Putnam, aims to show that natural kind terms acquire meaning in virtue of their reference (the real nature of those entities for which they stand). He believes that, in general, natural kind terms have their reference fixed by an initial baptism. 'Gold,' for example, might be defined as the substance which is instantiated by these items here.[12] Items which have similar superficial characteristics to those in the original set will then also be labelled 'gold'. But, with the advent of science, things change. Once the reference of the kind has been fixed, science determines what the real essence of the set is and so separates the genuine from the deviant members. After this, it becomes impossible for that kind not to have that

2: Objectivism

essence (for gold not to be the element with atomic number 79, for example) because *having that essence is just what it is to be a member of that kind.*

Kripke concocts many thought experiments to illustrate that, although reference-fixing tends to be a matter of superficial properties, still it is deeper, scientifically discoverable properties that determine what it is to be a member of a kind. Thus he suggests that we might discover that gold is in fact not yellow, but blue—we have all been deceived by some widespread optical illusion. In this case, he argues, it would not be announced that there is really no gold, but that although we supposed gold to be yellow, we were mistaken—it is actually blue. He also suggests that if we came across entities with the same surface characteristics, but quite different internal characteristics, from members of a particular natural kind, these entities would not belong to that kind because they would not possess the requisite nature or essence. We might thus discover some animals that look just like tigers, but which in fact have the internal structure of reptiles. These animals would not be tigers because they are not of the same species as tigers—they do not possess the same internal characteristics.

Like Putnam, Kripke advocates an objectivist epistemology to go alongside his objectivist metaphysics. Since human thought or knowledge does not influence natural kinds and essences, we can correctly speak of the essence of a kind without knowing what that essence is, he reasons:

> I think this is true of the concept of tiger *before* the internal structure of tigers has been investigated. Even though we don't *know* that internal structure of tigers, we suppose . . . that tigers form a certain species or natural kind. We then can imagine that there should be a creature which, though having all the external appearance of tigers, differs from them internally enough that we should say that it is not the same kind of thing. We can imagine it without knowing anything about this internal structure—what this internal structure is. We can say in advance that we use the term 'tiger' to designate a species, and that anything not of this species, even though it looks like a tiger, is not in fact a tiger.[13]

※

In this chapter, we have seen that a number of different philosophers have, over the centuries, embraced different forms or elements of Objectivism. We have learned that objectivist metaphysics speaks of an ordered, inherently structured, mind-independent world. We have also learned that, according to objectivist epistemology, accurate knowledge of that world involves reflecting its structure through impersonal, context-free eyes. Objectivism paints a tidy picture, but is that picture true? Is the world as straightforward as objectivists would have us believe? And is it reasonable to suggest that in experiencing the world, we simply slough off our humanity and observe through God's eyes?

3
Internal Realism I

In what ways does Objectivism fail? Two lines of attack lie open to us. We can argue that objectivist metaphysics is wrong and we can argue that objectivist epistemology is wrong. In this chapter, I concentrate largely on the failings of objectivist epistemology, introducing Hilary Putnam's internal realism as an alternative to Objectivism—an alternative that keeps a firm grasp on reality, while allowing that knowledge of that reality arises always from an internal, human perspective.

Countering objectivist *metaphysics*, however, involves countering the notion of an independently structured world consisting of entities, properties, relations and categories. This is exactly what I go on to do in chapter 7, where I suggest an alternative to the classical view of categorization, specifically in relation to natural kinds. My strategy involves illustrating that the way in which scientists categorize objects in the natural world is not purely a matter of the way the world is. Rather, scientists must make a *choice* as to which properties are to count as salient for matters of categorization. Which properties they choose will determine what counts as a relevant similarity for grouping entities together for categorization purposes. The underlying notion here is one of a world which is conceptually extremely rich, a world which can support more than one classification of its entities. Different scientists will highlight and downplay different aspects of reality and they can do this by taking different properties of real-world entities as salient for matters of classification. Which properties are judged essential for any one class will therefore depend, to a certain extent, on the aims and purposes of the scientist doing the classification (and so may vary, dependent upon those aims and purposes). And, of course, which properties are taken as essential dictates where the boundaries of the class fall. On this picture, members of so-called natural kinds, together with the essences and boundaries of these classes, become not purely a matter of metaphysics, but involve some degree of human input (epistemology). This picture casts doubt over the validity of objectivist metaphysics.

But what of objectivist epistemology? Hilary Putnam (in his later work[1]) provides us with a means of attacking the objectivist account of cognition, knowledge and meaning. As his arguments progress, we will see that objectivist metaphysics and epistemology are in fact closely interwoven. Rather than simply reflecting the world's independent structure, we humans actually contribute to that structure by means of the conceptual schemes which we impose upon the world.

Putnam aims to go beyond the philosophical positions of Subjectivism and Objectivism, to formulate a new position lying between these two extremes, a position which takes the best elements from each. He begins by characterizing "metaphysical realism"—another name for Objectivism. "The world consists of some fixed totality of mind-independent objects," he says. "There is exactly one true and complete description of 'the way the world is.' Truth involves some sort of correspondence relation between words or thought-signs and external things and sets of things."[2] This position—the "externalist perspective"—implies the existence of a God's Eye point of view, a way in which we can stand outside the universe and offer the uniquely correct description of it.

In opposition to metaphysical realism, Putnam launches an alternative view—internal realism—the core of which states that "*what objects does the world consist of?* is a question that it only makes sense to ask *within* a theory or description."[3] Talk of a God's Eye perspective is meaningless, since no such perspective can exist. Given a particular theory about, or description of, the world as a basic starting point, it is possible to talk about the way the world is, but without some kind of prior perspective, it is just impossible to talk.[4]

Truth, for the internal realist, becomes "some sort of (idealized) rational acceptability—some sort of ideal coherence of our beliefs with each other and with our experiences *as those experiences are themselves represented in our belief system.*"[5] Away goes the notion of correspondence between our experience and the world as it is in itself, independent of human cognition and language. If a God's Eye perspective is impossible to attain, then a comparison between our experience and an independent world also becomes impossible, since we can have no access to any such independent world. Our experiences, by definition, include something of us, as human beings. This something comprises a mixture of our physiology plus our beliefs, training and cultural inheritance—ingredients which have been inculcated in us since birth and which shape the way we cognize, think and speak about the world.

Does internal realism, then, lead us towards an out-and-out relativism, towards a situation where any conceptual scheme goes and where truth and reality become meaningless? No. As the name implies, internal realism still has its feet firmly in *reality*. It insists that we live in and interact with the world, constantly receiving cognitive inputs from it. Yet it also emphasises that those inputs are never direct, they are always mediated. Our experiences, knowledge, theories and descriptions are continually constrained by the world, but they are constrained by the world as *we* experience it.

In the absence of a unique, independent description of reality, the possibility that there can be more than one true theory or description of the world arises. There may exist "various points of view of actual persons reflecting various interests and purposes that their descriptions and theories subserve."[6] Linguist George Lakoff provides a useful illustration of this point.[7] There are various ways in which we can correctly describe a physical object like a chair. From the molecular perspective, it is a collection of molecules and not a "single undifferentiated bounded entity." From the perspective of wave equations in physics, there is no chair, only wave forms. From a lay human perspective, it is simply a chair—a single bounded entity. In each case, something exists in the external world, but that something can be described in very different ways from different perspectives or from within different conceptual schemes. It is only from the lay human perspective that the chair is considered a single object—even what constitutes an object and how the world should be divided into objects varies between theories of description. Each perspective is correct, but it is correct from *within* the relevant conceptual scheme or theory of description.

For the internal realist, then, meaning cannot be a matter of a word or concept correctly referring to some entity in an objectivist world (since we do not have experience of the objectivist world). Rather, meaning is a matter of a community of language users using signs in a particular way to refer to entities *as those language users experience those entities*. Putnam believes that, "'objects' do not exist independently of conceptual schemes. We cut up the world into objects when we introduce one or another scheme of description.

Since the objects and the signs are alike *internal* to the scheme of description, it is possible to say what is what."[8]

Likewise with kinds or classes of objects. The objectivist argues that the world in itself falls into entities, which themselves fall into classes. To say that something is of the same kind as something else is therefore to say that really, objectively or metaphysically, the two things belong together in the same class. Yet the internal realist has shown that it is meaningless to talk of direct access to the world in this way. "'Of the same kind,'" therefore, "makes no sense apart from a categorial system which says what properties do and what properties do not count as similarities."[9] When we cut the world up into objects, we also cut it up into kinds. By isolating, defining and labeling an object, we *ipso facto* designate the kind to which that object belongs. All other objects that we judge relevantly similar to that object will belong to the same kind. And this is a matter not simply of reflecting the inherent structure of an objectivist world, but of interpreting and shaping up our irreducibly human experiences. It is a matter of deciding in what respects two things must be similar in order to belong to the same class—a decision which is made relative to a conceptual scheme or theory of description which prefers certain properties over others for the purposes of classification.

TRUTH AND INTERNAL REALISM

The insights of internal realism help us to appreciate some of the shortcomings of Objectivism. It is impossible to view the world through an impersonal, context-free lens. Experience presupposes an entire biological, cultural and personal background. We impose structure on the world as we experience it. The world can support differing descriptions and classifications of the entities within it. And so on. Yet one of the issues that Putnam discusses—truth—deserves closer critical scrutiny. Although he aims to introduce a new theory of truth to counter the traditional correspondence theory of truth (which tends to go hand in hand with metaphysical realism), his account is confused and unnecessarily complex. At times, it even smacks of the metaphysical realism he claims to have jettisoned. Can we provide a more adequate conception of truth without abandoning the basic tenets of internal realism?

Recall Putnam's claim that, on the internal realist picture, truth is "some sort of (idealised) rational acceptability—some sort of ideal coherence of our beliefs with each other and with our experiences *as those experiences are themselves represented in our belief system*—and not correspondence with mind-independent or discourse-independent 'states of affairs.'"[10] This, of course, makes sense, since, according to internal realism, the notion of a God's Eye, context-free point of view is meaningless. If our true statements accord with anything, then it must be with our experience of the world and not with a God's Eye notion of that world. As Putnam puts it, "You can't single out a correspondence between two things by just squeezing *one* of them hard (or doing anything else to just one of them); you cannot single out a correspondence between our concepts and the supposed noumenal objects without access to the noumenal objects."[11]

Putnam goes on to define rational acceptability, explaining that what makes a statement or a theory rationally acceptable is coherence of theoretical beliefs with each other and with experiential beliefs and coherence of experiential with theoretical beliefs. Our notions of acceptability and coherence are, as we have already seen, dependent on our humanity—our psychology, biology and inherited culture. And so it becomes possible to arrive at two theories or conceptual schemes which are equally coherent and rationally acceptable. It is possible to end up with more than one true description of reality. A consequence of this, according to Putnam, is that scientific truth depends upon certain values, since "*truth is not the bottom line:* truth itself gets its life from our criteria of rational acceptability, and these are what we must look at if we wish to discover the values which are really implicit in science."[12] The claim that science aims to construct a true picture of the world is really meaningless until one spells out the criteria of rational acceptability to which science adheres.

Yet Putnam insists that rejecting the metaphysical realist's conception of truth does not amount to *identifying* truth with rational acceptability—the two are not equivalent notions. What is rationally acceptable may change over time and what is rationally acceptable is a matter of degree. (We consider truth, by contrast, an absolute notion—a statement is either true or false, rather than true to a certain degree.) He therefore defines truth as an *idealization* of rational acceptability, saying, "We speak as if there were such things as epistemically ideal conditions, and we call a statement 'true' if it would be justified under such conditions."[13] He acknowledges that we can never really attain epistemically ideal conditions or even be sure that we come anywhere close to them. They function in the same way as frictionless planes, which cannot be attained, but are nevertheless useful to us since we can approximate them very closely. Putnam's implication seems to be that we can approximate ideal epistemic conditions very closely too.

In a later work—*Realism with a Human Face*—Putnam attempts to explicate the notion of epistemic conditions further. To claim that a statement is true in its context or conceptual scheme is to claim that it could be justified were epistemic conditions good enough, or ideal, he says. He provides the following example. An ideal epistemic condition in relation to Putnam's claim that there is a chair in his study would involve him being present in the room with the lights on or with daylight coming through the window, regarding the chair with unimpaired eyesight and a clear mind. It is even possible to drop the word 'ideal' altogether and simply say that there are better and worse epistemic situations with regard to particular statements, he claims. On this wording, the above example would represent a very good epistemic situation.

Despite the soundness of Putnam's basic intuitions—that correspondence with a God's Eye notion of the world is untenable and that truth must therefore rest on some kind of agreement between our theorizing about and our experience of the world—he seems to come unstuck over the notion of idealized rational acceptability. The crux of the problem lies in his insistence that idealized rational acceptability and ideal epistemic conditions are mere fictions and cannot be attained by us. This in turn suggests that we can never attain truth, that what we do attain cannot be counted as truth and that what we do have is in some way inferior to the unattainable ideal of truth. By holding on to this notion of truth, Putnam appears to be reinstating a situation of the very kind which he was aiming to reject. On the one hand, we have what is rationally acceptable for humans to believe—what we utilize in our negotiation of the world—and on the other, we have truth (idealized rational acceptability)—some kind of abstract notion which is unavailable for human use. Idealized rational acceptability appears to play a similar role to that played by the God's Eye view in metaphysical realism. Idealized rational acceptability seems to comprise a construct which is entirely independent of human cognition and judgement—a kind of view from nowhere. And such a view is surely at odds with the moderate, non-divisive picture Putnam claims to present. Idealized rational acceptability, I suggest, should be rejected along with, and for the same reasons as, the God's Eye view of the world.

Worse still, Putnam's two accounts contradict one another. The claim that there exist ideal but unattainable epistemic conditions is not at all the same as the claim that there exist better and worse epistemic conditions with regard to particular statements. No one would deny the latter pronouncement, and we would surely agree that a statement made in good epistemic conditions is more rationally acceptable (and so more likely to be true) than one made in poorer epistemic conditions. However, good or very good epistemic situations are ones which *are attainable by humans* while, by definition, unattainable or fictitious epistemic conditions are not. And Putnam's chair in the study example involves epistemic conditions which *are* possible to attain.

Given the inconsistencies in Putnam's account, we would do well to look elsewhere for an alternative conception of truth, a conception that remains simple yet preserves the basic intuitions of internal realism. Philosopher Mark Johnson provides a promising alternative.

In his book, *The Body in the Mind*, Johnson develops Putnam's internal realism further, with special reference to its implications for a theory of meaning. He stresses the human nature of meaning, arguing that linguistic items do not acquire meaning by latching onto and mirroring the inherent logical structure of the world, as Objectivism would have it. Rather, they acquire meaning by being used by humans in certain ways for certain purposes. Linguistic symbols reflect human *understanding* of the world, he insists. Despite the stock of shared assumptions between these two thinkers, Johnson retains, in relation to truth, a realist intuition which Putnam abandons. Johnson preserves the notion that truth involves statements which *correspond* to states-of-affairs in the world.

"Truth-as-correspondence is still a workable notion only if it is not understood in the objectivist fashion, as requiring a God's-Eye-View of an external relation between words and the world," Johnson claims.[14] He argues that, given a particular conceptual scheme which divides the world into entities and kinds of entity, certain statements will correspond more accurately to that world than others. But he qualifies this statement: "'Correspondence' will always be relative to our *understanding* of our world (or present situation) and of the words we use to describe it."[15] Statements cannot simply correspond to states-of-affairs in themselves in one and only one way, since this kind of unmediated, God's Eye mapping onto the world is impossible, as Putnam has shown. Rather, we employ statements for particular purposes—statements which correspond to our experience and understanding of the world in which we are situated. Since correspondence is not absolute, it becomes feasible that different statements correspond in different ways, dependent on the aims and purposes of the people who make those statements and who have the experience to which they correspond.

We do not require an absolute, God's Eye notion of truth because we view the world "through shared, public eyes" which are a product of "our embodiment, our history, our culture, our language, our institutions." Our shared, public vision ensures that we stay in tune with reality "not in the 'one correct way' but in one or more of the possible ways in which Nature can be described." In this way we can "still preserve a notion of truth-as-correspondence, as long as it is contextually situated."[16]

With the God's Eye notion of the world lying in tatters, it makes no sense to define truth in an absolute way. Yet, given a particular context, conceptual scheme or point of view on the world, it makes sense to differentiate between true and false statements. As we will see in future chapters, Johnson's account of truth fits well with my pluralist account of scientific classification. My suggestions—that differing theories lead to competing definitions of kinds and to competing classifications of particular entities—lead to a situation where competing definitions and statements can all be true, just so long as they are equally scientific and fit (our experience of) the facts equally well. My account of scientific classification concurs with Johnson's view that

> "Accurately describing reality" is not a single, homogeneous purpose on a par with a purpose like making one's bed. "Describing accurately how things are" is a shorthand for "finding descriptions of reality that work more or less well given our purposes in framing descriptions of reality." ... Truth is always truth relative to a basic description and relative to standards of adequacy determined by our human purposes and the nature of our interactions with our environment.[17]

Putnam's internal realism combined with Johnson's contextual correspondence theory of truth provides a powerful weapon against Objectivism. Objectivism's weaknesses—the incoherence of the God's

Eye view and the lack of regard for any human perspective—provide a launching pad for internal realism. We have learned that knowledge of the world can never be pure, direct or unmediated, since experience—and so knowledge—requires prior conceptualization or structuring; it requires the adoption of a perspective. And we have seen that a perspectival approach allows for more than one true description of the world. We have also learned that acknowledging the human perspective does not lead to an 'anything goes' relativism. We live in and interact with the world and this world constrains our experience, our descriptions and our truth claims. Internal realism reasserts the role played by the human agent in matters of knowledge, truth, meaning and language, but it does so without detracting from the reality of the external world.

4
Internal Realism II: Criticisms and Implications

Internal realism, of course, has its critics. Since I adopt this position as a theoretical backdrop for my account of scientific classification, I need to tackle these criticisms. By doing this, I further strengthen the case against Objectivism, as well as providing additional elucidation of internal realism. Having dealt with some common criticisms of internal realism, I move on to examine how this doctrine complements my pluralist account of scientific classification and to discuss how my account relates to notions of objectivity and realism propounded by contemporary philosophers of science.

Hartry Field is representative in raising a number of common objections to internal realism.

Objection 1. Field correctly characterizes Putnam's account of metaphysical realism with the following three statements:

- Metaphysical realism$_1$ = "the world consists of some fixed totality of mind-independent objects"
- Metaphysical realism$_2$ = "there is exactly one true and complete description of 'the way the world is'"
- Metaphysical realism$_3$ = "truth involves some sort of correspondence relation between words or thought signs and external things and sets of things"[1]

He then proceeds to argue that metaphysical realism$_2$ is not a consequence of metaphysical realism$_1$, that metaphysical realism$_2$ is false and that it "should not be taken as a component of any sane version of realism."[2] It would be possible, he claims, for there to be a true description of the world in a language completely alien to our own—it might, for example, use predicates with extensions which are not easily definable or not definable at all in our own language or it might even use referential mechanisms which are quite different from predicates. How can we say that two such true, yet different, descriptions of the world are the same?

In order to get a response to Field off the ground, we need to reassure ourselves that proponents of Objectivism or metaphysical realism *do* take on board the notion that there is "exactly one true and complete description" of the world. Recall those philosophers considered in chapter 2.

For Aristotle, the world is divided inherently into species and genera, and giving a correct classification of all the entities in the natural world involves reflecting that natural order. Certain remarks imply that each entity possesses just one correct classificatory position, hence one unique essence: "the essence is what something is" and "each thing and its essence are one and the same in no merely accidental way."[3] It therefore seems reasonable to infer that this unique reflection of the natural order comprises the one true and complete description of the world. In this case, completeness, for Aristotle, involves accommodating or placing all entities within his 'scientific' (informed by nature) classificatory scheme. And this, of course, involves a prior judgement concerning the structure of the scheme and those properties to be used when placing things within it. Completeness, then, does not entail describing all entities in every possible way, but instead entails classifying all entities within a scheme judged better or more adequate than other possible schemes.

Kripke and (the early) Putnam constantly remind us that a member of a particular natural kind cannot fail to possess the kind's essence since possession of that essence is simply what it is to be a member of that kind. Consequently, we would be *wrong* to place an entity in a natural kind if that entity lacked the requisite essence. For Kripke and Putnam, the one true and complete description of the kinds and essences of the world is the one which reflects these kinds and essences, as they exist in nature. Any other description is false. Kripke and Putnam therefore share in the Aristotelian notion of completeness—in order to arrive at a true and complete description of the (natural) world, we ought not to describe all entities in all possible ways, but should instead classify all entities on the basis of their essential properties, as uncovered by science. The classifications offered by science comprise a complete description of the world since, in Kripke and Putnam's eyes, science provides the superlative description—the one that most accurately describes nature. Aristotle, Kripke and Putnam all appear, then, to equate completeness with a certain definitiveness that they ascribe to science on the grounds that science correctly reflects the natural order.

Wittgenstein argues that language represents the way things are in the world and so comprises propositions which correspond isomorphically to facts and which share the same structure as those facts. The one true and complete description of the way the world is involves, for Wittgenstein, a series of propositions which correspond exactly to the totality of facts which comprise that world. He even says, "If all true elementary propositions are given, the result is a complete description of the world."[4]

A number of metaphysical realists therefore *do* believe that there is a unique, complete and true description of the way the world is.[5] Whether this position is sane or insane, true or false, is quite another question, but it is a position which has been held and so (the later) Putnam can quite legitimately argue against it.

Putnam provides his own response to Field's criticism. One of the ways in which we might make sense of the claim that there is one true and complete description of the world is to assume that metaphysical realism$_1$ is true. In this case, there will be a definite set of individuals, *I*, of which the world consists, and a definite set of all properties and relations which pertain to those individuals, *P*. We can then imagine an ideal language which has a name for each entity in *I* and a predicate for every member of *P*. As a result of isomorphism with the world, this language and its set of true sentences can be construed as the unique, complete and true description of that world. Thus metaphysical realism$_1$ and metaphysical realism$_2$ are mutually supporting doctrines.

We can understand Putnam's point using a slightly different description. If a metaphysical realist accepts that the world consists of a fixed totality of mind-independent objects and also accepts that truth involves some sort of correspondence between symbols and external things, then it would seem to follow that this metaphysical realist accepts the notion that there is one complete and true description of reality. To give a complete and true description of the world is to give a list of *all* the statements which are true of the world in virtue of corresponding to *each and every* element of, or state-of-affairs in, that world. But in this case, Putnam's definition appears too broad to cover certain versions of metaphysical realism. As we have already seen with Aristotle, Kripke and Putnam's earlier work, description can involve some kind of prior selection and, in this situation, completeness involves giving a complete description from a particular (supposedly definitive) point of view, rather than providing complete descriptions from all possible points of view.

What, then, of Field's claim that concepts very different from our own might be used to describe the world? It looks, at this point, as if Field is buying into one of the tenets of *internal* realism—that the world can admit of more than one true or correct description—rather than producing a criticism of metaphysical realism, as characterized by Putnam. If he wants to pursue this line of attack any further, then, he must provide stronger arguments concerning (a) why he disagrees with metaphysical realism$_2$ and why metaphysical realism$_2$ is not entailed by metaphysical realism$_1$, and (b) why he is not prepared to accept Putnam's internal realism.

Objection 2. Field takes issue with Putnam's claim that "*what objects does the world consist of?* is a question that it only makes sense to ask *within* a theory or description,"[6] correctly assuming that Putnam means that "different correct theories or descriptions will answer the question differently."[7] He continues by attacking Putnam's suggestion that Maxwell's electromagnetic field theory and action-at-a-distance formulations of electrodynamics are both equally good and coherent theories which fit our experiential beliefs equally well, despite the fact that they are metaphysically incompatible. (One claims that 'fields' exist and the other claims that they do not.) Field argues that this example neither suggests that objects are mind- or theory-dependent, nor does it pose a problem for the correspondence theory of truth. Advocates of field theory need not deny that the behavior of particles can be explained without fields, just so long as we "introduce enough complexity into the basic equations of motion."[8] Likewise, advocates of action-at-a-distance theories need not deny the existence of fields, but "merely refrain from asserting their existence."[9] "Most of us recognise that more exists than one need assert the existence of: it is rarely to the point to assert the existence of undetached rabbit parts as well as of rabbits . . . there is clearly no way to draw anti-realist consequences merely from the fact that two equally good theories could differ in their existence claims," he concludes.[10]

We can attack Field on several fronts here. First, note that Putnam claims that these two theories are *metaphysically* incompatible, meaning that the idea that both theories can be correct is a problem for the *metaphysical* realist. Recall that the metaphysical realist—as characterized by Putnam—believes that (a) there is a fixed mind-independent totality of objects, (b) there is exactly one true and complete description or theory of the world and (c) truth involves a correspondence between words or thoughts and external things and sets of things. Clearly, then, this example *is* problematic for Putnam's metaphysical realist. By dint of (a), either fields must exist or they must not, yet the very point of this example is that we cannot answer conclusively—whether fields exist or not will depend on which theory we are working within. By dint of (b), there cannot be more than one theory which correctly describes the same phenomenon, yet in this example, we have two theories which explain the same phenomenon equally well.

So, Field's claims that more exists than need be asserted will be of small comfort to Putnam's metaphysical realist, since this is just what he believes to be impossible. If the world consists of a fixed totality of objects and if there is just one true theory of the world, then all that needs to be asserted (by the true theory) will be all that *can* exist. Conflicting theories and incompatible objects are just not acceptable to the metaphysical realist. Field has, of course, previously stated that he disagrees with Putnam's characterization of metaphysical realism, but (a) we have already illustrated that his disagreement is unfounded and (b) *given* Putnam's characterization, the field/particle example *is* problematic for the metaphysical realist.

This example also poses a problem for the correspondence theory of truth. Given that the world consists of a fixed totality of individuals plus properties and relations pertaining to those individuals, what makes a word or thought or statement true is that it corresponds one-to-one with individuals/properties/relations in that world. If the statement, "physical events = particles acting at a distance" is true, then it must be a fact of the external world that there exist particles which act at a distance. However, if there are fields in the external world which mediate the action of these particles, then this statement will be false. If the objects of the world are a fixed set and there is only one true description of that world and truth involves unique correspondence to the fixed objects of the world, then both states-of-affairs cannot simultaneously exist. Field's attempt to fix the problem in terms of what we can assert simply does not work on Putnam's characterization of metaphysical realism.

Yet, for the internal realist, these two incompatible theories do not pose a problem. In fact, Field's notion that more exists than we need assert fits the internal realist picture very well. As Putnam himself explains, "[it] is not that correspondences between words or concepts and other entities don't exist, but that *too many* correspondences exist."[11] Without the restriction of a God's Eye point of view, a pluralist conception of truth becomes possible. Internal realism therefore very much allows for the suggestion that more exists than we need assert—it allows for the fact that fields might exist, yet whether or not we assert their existence will depend upon which theory we are working within.

We should perhaps also say something about Field's insistence that anti-realist consequences cannot be drawn from the field/action-at-a-distance example. Remember that Putnam advocates a position called internal *realism*, insisting that we humans live in, and receive continuous experiential inputs from, reality, but he denies that these experiences come from a God's Eye point of view. He does not, then, draw anti-realist—but anti-*metaphysical* realist—consequences from the field/action-at-a-distance example. By conflating these two quite different projects, Field entirely overlooks the insights of Putnam's position.

Objection 3. Field interprets Putnam's statements that "objects do not exist independently of our conceptual scheme"[12] and "objects themselves are as much made as discovered"[13] to mean that we might have had conceptual schemes so different that we did not think in terms of, for example, dinosaurs, but in terms of quite different entities. Field offers undetached dinosaur parts as a suggestion. But this is simply not enough to make Putnam's metaphor work, he complains, reiterating the point that more exists than we need assert—both the dinosaur and the undetached dinosaur parts exist, whether anyone asserts that they exist or not, whether anyone thinks in terms of them or not. Furthermore, he is concerned that as the book goes on, "one becomes more and more tempted to think that Putnam does in some literal sense believe that we created the dinosaurs. Again and again . . . he tells us that facts depend on human cognitive values."[14]

Field seems not to have fully understood Putnam here. According to Putnam, yes we do 'make' objects, but of course not literally. The best way to understand this point is in terms of kinds. Objects do not exist independently of conceptual schemes because it is only under a particular conceptual scheme that two things can be identified as being of the same kind (of object). To pick out and label an object is to associate particular properties with that object in virtue of which other objects of the same kind can be identified. And, since the world does not come pre-formatted with regard to salient properties, our conceptual schemes do the job of isolating specific properties which we then use in judging objects as similar and belonging to the same kind.

Coincidentally, one of the case studies we will discuss in chapter 7 explains how one man—Richard Owen—'created' the category 'dinosaur'. Prior to Owen, what we now call dinosaurs were viewed as enormous reptiles—'fossil lizards'—based on fossil remains of jaws and teeth which indicated similarities to extant lizards. Yet Owen gave these creatures their own taxonomic rank—*Dinosauria*—based on anatomical peculiarities of the sacrum, ribs and extremities and on their enormous size, which served to distinguish them from living lizards as well as from Mesozoic marine lizards. Here we have an example of an object being 'made' as much as discovered. Whether something counts as a dinosaur or not will depend on whether you are working within a conceptual scheme which counts dinosaurs as a separate class of things. If it does, then there will be properties associated with that class in virtue of which objects can be identified as instances of the class. If not, then the objects which we call dinosaurs will fall in with some other class on the basis of a different set of salient properties.

Of course we do not literally make dinosaurs. Dinosaurs—those particular bits of matter exhibiting those particular properties—existed at some point in the past no matter what we think or say. However, the fact *that* something existed does not necessarily wholly determine exactly *what* existed. Humans have had a hand in making these entities *dinosaurs*, as opposed to giant lizards. When Putnam speaks of facts depending on human cognitive values, then, he is not arguing that we literally manufactured the dinosaurs, but that how we divide up and categorize our environment depends, at least partially, upon properties and factors we take to be salient, on what is of interest *for us*.

Yet Field's observation that this is not nearly enough to make Putnam's metaphor work is, perhaps, partially justified. Since the verb 'to make' tends to carry connotations of physically constructing, it was probably a poor choice of term on Putnam's part. And Putnam seems subsequently to have realized this—in more recent work, he expresses regret at having talked about the mind-dependence of the world, while continuing to assert that a variety of competing descriptions may all be true of the world.[15] Even so, we must guard against taking sentences out of context and criticizing in isolation. The overall thrust of Putnam's position remains that our experience and description of the world involve all sorts of intrinsically human predispositions, interests and choices—a picture which the *metaphysical* realist is at pains to avoid.

Having defused some of the more common objections to internal realism, we are now in a position to consider how this doctrine fits with the aims of, and so acts as a backdrop for, my account of scientific classification.

As we have already seen, I aim to show that scientific classification of natural (non-manufactured) objects involves not simply a reflection of the way the world is (metaphysics), but also an element of choice or decision on the part of the scientist doing the classifying (epistemology). Putnam's internal realism enables us to understand how such a non-objectivist account of scientific classification is plausible. When scientists classify, they operate with an *internal* perspective. They cannot stand outside the world, or operate with an external perspective, because classification (like description) involves interaction with and participation in the world, it involves the adoption of a perspective. In choosing to adopt a particular perspective, a scientist embraces a conceptual scheme that isolates which of the many available properties do, and which do not, count for the purposes of classification.

This does not, of course, mean that scientists may classify in any way that takes their fancy. The environment exhibits its own characteristics and so places limits on categorizations of the entities within it. Ignoring these limits may cost us dearly. No amount of scientific conceptualizing can alter the fact that some plants are poisonous to humans and others are not, for example. My argument is simply that the world does not come with one preferred theory of description.

We therefore naturally build a description of the world which complements our aims and purposes—and different aims and purposes may result in differing descriptions.

Since scientists must adopt an internal perspective on the real world, I suggest that scientific classification involves a *synthesis of metaphysics and epistemology*. The specific perspective a scientist adopts (his/her epistemology) governs which aspects of the real world (the metaphysics) his/her classifications will reflect. Without embracing some conceptual scheme—without activating an epistemology—description of the way the world is cannot commence. I concur with internal realism's move away from the split human/world scenario invoked by both Objectivism and Subjectivism. I concur with internal realism's presupposition that we form a *part* of the environment in which we operate. (Human) organisms and the environment affect one another, causality and influence working in both directions. Through our continuous interaction with the environment, we learn how to negotiate and describe its limits in ways which best suit our own purposes.

The tenets of internal realism also fit snugly with my use of empirical evidence in future chapters. Chapter 5 explores the development of the psychology of categorization and emphasizes that the majority of human concepts simply do not fit a classical-type structure. In chapter 7 I use material from contemporary biology and the history and sociology of science to promote my pluralistic understanding of scientific classification.

The objectivist, of course, will be uninterested in empirical evidence concerning our concepts or the means by which we classify entities. What people *actually do* is of no relevance to his/her theory. All s/he requires is right there in the world—human vagueness and ambiguity simply interfere with the clarity and precision of the metaphysics. The objectivist may, perhaps, show some interest in those situations where we 'get it right', where we 'correctly' mirror the structure of the world, but still this is parasitic upon metaphysics, since it is but a reflection of the metaphysics. The internal realist, by contrast, will consider empirical material relevant, since s/he considers that, through thought and conceptualization, we have a hand in shaping the environment around us. Epistemology is, for the internal realist, relevant to metaphysics. If psychological evidence tells us that the majority of human concepts do not exhibit a classical-type structure, the internal realist will accept this as a statement about the way in which we impose structure upon our world. S/he will also accept this as telling us something about the way the world is. And, if case studies suggest that human aims, theories and commitments play a decisive role in scientific classification, the internal realist will not show surprise, since this is just what internal realism leads us to expect.

My account of classification is, of course, making claims about *science*—about how scientists operate within, cognize and describe the natural world—whereas Putnam's internal realism deals with human cognition and description in general—it is not restricted to human scientists. How, then, does my work relate to that of other philosophers of science? Two contemporary philosophers of science—Helen Longino and Ronald Giere—aim (as I do) to sketch a middle path between the extremes of Objectivism and Subjectivism. More precisely, they aim to produce accounts of scientific knowledge and practice that form a third way between the rigid rationality favored by positivist pictures of science and the inherent relativism entailed by recent theorizing in the sociology and history of science. They want, as do I, to maintain that the practice of science remains a realistic and rational business, while acknowledging that science is also an inescapably *human* activity in which scientists interpret and make decisions about the world they study.

Longino promotes a contextualist model of science, arguing that science is practiced within a social and cultural environment, yet assuring that social and cultural values do not destroy the possibility of objectivity. In fact, it is this social character of science that ensures its objectivity. "What I wish particularly to stress," she says, "is that the objectivity of scientific inquiry is a consequence of this inquiry's being a social, and not an individual enterprise."[16] Scientific knowledge is produced by many individual scientists working together—both officially as part of the same research group and more generally in the sense of investigating common problems. (Parts of) theories and models are produced by different individuals; theories and hypotheses are criticized and modified by people other than their originators; experiments are repeated and altered by various scientists; scientific activity is funded and its results published only following a lengthy process of peer review. Science, then, is a collective, social activity.

The social character of science ensures its publicity, Longino continues, and more particularly, the possibility of intersubjective criticism ensures objectivity, despite science's contextual nature. One type of criticism is particularly important for ensuring objectivity—the type that questions "the background beliefs or assumptions in the light of which states of affairs become evidence [in support of hypotheses]."[17] So long as these background beliefs remain subject to criticism from the community of scientists, they can be defended, modified or abandoned as required. In this way, the welcoming of hypotheses into the sanctum of scientific knowledge remains free of individual, subjective interest and preference. Longino concludes: "Only if the products of inquiry are understood to be formed by the kind of critical discussion that is possible among a plurality of individuals about a commonly accessible phenomenon, can we see how they count as knowledge rather than opinion."[18]

Longino's account of science as social knowledge establishes points with which I agree. She presses the human nature of scientific activity and, as we have already seen, I argue that the process of scientific classification involves adopting a human perspective on the world. Yet her angle of attack is different—she criticizes the notion that science is an activity practiced by individuals rather than social groups, whereas I attack the possibility of a universal, context-free point of view on the world. I certainly do not deny that science is a social activity—one of the reasons for the success of a particular scheme of classification, for example, will be that it is accepted by (a portion of) the scientific community. But—and this leads on to a second difference in our views—acceptance by the scientific community alone will not guarantee the success of the scheme. The scheme must also reflect some scientifically significant pattern of similarity that really does occur in the natural environment. And, I suggest, it is this latter feature which underlies acceptance of the scheme by the wider scientific community. Longino and I both agree that science as a human activity does not entail relativism, but the standards against which we judge the rationality of science are different. For me, the primary relevant standard comprises regularities occurring in the real world. The human element is, I argue, entailed by the fact that the real world is simply too complex to admit of a unique, all-purpose description or scheme of classification. For Longino, the primary relevant standard comprises the possibility of criticism by, and so ultimately agreement within, the scientific community.

4: Internal Realism II

Giere, unlike Longino, does not seek to establish that science occurs in a social context, but takes this as given. "Obviously, the acquisition of scientific knowledge, like all human activities, takes place in a social environment," he says. "That is not an issue."[19] Rather, he wants to understand the success of science in relation to the fact that science takes place in a social environment and that we are biological creatures who have evolved the cognitive capacity to interact with the world surrounding us. He aims to fit these various elements together without relying either on traditional philosophical theories of science which depict scientists as ideally rational (and so fail to account for widely divergent opinions in science) or on social constructionist approaches which depict scientific 'knowledge' arising from competing, vested professional and social interests (and so fail to explain the evident success of science and technology).

Giere understands scientific theories as internal representations of the external world. He is therefore committed to realism, but does not accept that the human mind mirrors nature, nor that we can have access to the world independent of internal, cognitive processing. He thus sees scientific theories (or representations) as "fitting the world in limited respects and degrees, and for various purposes."[20]

Constructive realism is Giere's formal attempt to portray a version of scientific realism which pays tribute to the external world while doing justice to the role of the scientist as a biological and social agent. Scientists construct theoretical models which they intend as representations of (aspects of) the real world. The primary relationship between these models and the world is similarity, Giere argues, and since things can bear similarity to other things in a multiplicity of ways, the similarity relation must be limited in both respect and degree. Wherein does the constructivism lie? Giere emphasizes that models are "deliberately created" by scientists, saying, "Nature does not reveal to us directly how best to represent her. I see no reason why realists should not also enjoy this insight."[21] He thus denies Objectivism's uniqueness thesis. Furthermore, constructive realism is a restricted version of realism, since theoretical hypotheses are understood as asserting similarity between a real-world system and some—but not necessarily all—aspects of a model. Scientists themselves must decide in particular cases which aspects are and which are not to be counted as similar, and why.

Giere's account of the scientific enterprise comes closer to my views than does Longino's. He, like me, confirms that scientists remain in touch with the real world, but denies that scientific knowledge can therefore be impartial or context-free. Scientists are, after all, humans—sophisticated products of the evolutionary process—and their cognition and understanding of the world will necessarily reflect their biology, as does all human experience and understanding. Interestingly, he claims that, despite superficial differences, models used by scientists and those used by laypeople are "fundamentally . . . the same sort of thing."[22] In chapter 7, I make a similar claim regarding scientific classification, illustrating that the psychological explanation-based model of categorization applies not only to classification conducted by laypeople, but also to classification conducted by scientific experts. Scientific classification, I suggest, is much more similar to lay classification than traditional (objectivist) accounts imply.

Finally, Giere insists, as do I, that scientists make a choice when representing the external world. Since the world does not come with a preferred (or best) description, scientists must decide what kind of models to use for representation as well as deciding which aspects of a given model bear similarity to (parts of) the real world. Giere and I both allow that the environment can support more than one representation of itself, which in turn involves scientists in decision-making as well as reflection. Realism and construction, on both our accounts, go hand-in-hand.

5

The Psychology of Categorization

It is not only philosophers who study the nature of classification. There is also an important area of psychology, known as the psychology of categorization, which examines the way in which we sort and categorize the objects we experience, the aim being to clarify how we mentally represent concepts. The psychology of categorization, however, (unlike, for example, the Putnam/Kripke approach to natural kinds) examines the way in which laypeople—and not scientists—categorize objects.

One of my own aims, as stated in chapter 1, is to take one of the most recent psychological theories of categorization—the explanation-based view—and illustrate how its basic principles can be applied to classification carried out by scientific experts as well as laypeople. In order to set the stage for this, we need to explore the history of the psychology of categorization. As we do this, we will see that there has been a shift away from tightly-defined accounts (the classical view) towards much more flexible, human-oriented accounts (the explanation-based view), a shift that mirrors the movement in philosophy away from Objectivism and towards internal realism.

THE CLASSICAL VIEW

Until the mid-1970s, the classical view of categorization was accepted wholesale and without question in psychology. The key notion here, as we saw in chapter 2, is that of necessary and sufficient conditions. Each of our concepts has a set of defining conditions (properties) attached to it, which are individually necessary and jointly sufficient for membership in that class. To decide whether an object belongs to a particular class, all one need do is ascertain whether it satisfies the defining conditions attached to that class. We might, for example, claim that the defining conditions for membership in the class of chairs comprise having a seat, a back, four legs and being made of wood. In this case, an object possessing all four features would qualify as a chair, while an object lacking one or more of these features would not.

The classical view has some pretty restrictive consequences. It entails that membership in a class is an all-or-none affair. Either something belongs to a class or it does not. There is no question of fuzzy boundaries between classes, with queries as to which category an object falls into. Objects belong properly in one class only. Each member of the category is as much a member as the next; no provision is made for some objects being perceived as more central or typical members of the category than others. This means that in order to judge that an object belongs to a particular class, we simply require definitional knowledge. All we need are the defining conditions attached to the class and the ability to decide whether that object exhibits those particular features. General world knowledge (knowledge of how the extension of a concept relates to or interacts with the extensions of other concepts) apparently plays no part in this cognitive process. The classical view, as we saw in chapter 2, is a thoroughly objectivist model.

During the 1970s, several psychologists—most notably Eleanor Rosch and her colleagues—began to call into question, and produce convincing experimental evidence against, the classical view.

Perhaps the major problem for the classical view lies in the fact that the categories we use clearly *do* have fuzzy boundaries. Class membership is not an all-or-none affair and it can be unclear whether an object belongs in one category or in another. Do rugs and television sets belong to the class of furniture? Is a hammock a chair or a bed? Is chess a game or a sport?

Michael McCloskey and Sam Glucksberg, among others, built a psychological experiment around this intuition. They asked their subjects to judge whether particular objects belonged to certain familiar categories. These objects had antecedently been judged for degree of typicality in the class by an independent group of subjects. The results show that for those objects judged to be of only an intermediate level of typicality (e.g., bookends = furniture), subjects frequently disagreed with one another as to whether the objects belonged in the category or not. Further, after a one-month period, it was found that subjects showed a tendency to change their own decisions about category membership. In contrast, with objects judged to be highly typical of a category (e.g., chair = furniture) and with those objects which were totally unrelated to the category (e.g., cucumber = furniture), subjects agreed with each other and were consistent in their own judgements.[1]

These results strongly suggest that categories have fuzzy boundaries. If membership were all-or-none, there should be no disagreement, either inter- or intra-subject, regarding that membership. The boundaries between classes should be clear-cut, with members and non-members easily distinguishable, without reservation.

A second problem for the classical view is that our categories exhibit prototypicality, which results in certain typicality effects. It seems that as part of everyday life we quite naturally distinguish between typical and atypical members of a class. Said of a beanbag, for example, "That's a funny sort of chair!" Or of the merits of apples versus lychees, "They are proper fruit. I don't like these new-fangled, exotic things on sale these days!"

This intuition has again been exploited in psychological research. There exists overwhelming experimental evidence that subjects will readily judge certain members of a category to be better examples, or more typical of the category, than others.[2] This appears perfectly natural and being asked to judge typicality produces no surprise or difficulty. Furthermore, this tendency has been shown to have effects in relation to other cognitive tasks. When subjects are asked to list instances of a concept, the more typical exemplars are listed first; when asked whether an object belongs to a category, the more typical the object is of the category, the quicker a response is elicited from a subject; children tend to learn highly typical exemplars of a category before less typical exemplars; and so on.

One final problem for the classical view is that, for most natural concepts, no one has yet come up with the requisite sets of necessary and sufficient defining conditions. No sooner do we come up with a set, than we think of instances of the concept which do not satisfy those conditions. It is likely, for instance, that a beanbag will fail to meet any of the conditions which we might expect to find attached to the concept 'chair.' It seems unlikely that a tiger which lost its stripes or a leopard which lost its spots would cease to be considered a tiger or a leopard by us. Until very recently, books were only printed on paper, but now we can read them on-line.

Empirical evidence supports this intuition. Francis Bellezza asked a number of subjects to define certain commonly used nouns. He then repeated the exact same experiment one week later. Results showed that different subjects varied considerably in the features which they used to define these nouns. Furthermore, in the repeat experiment, it was found that individual subjects defined the same nouns differently—some of the features they used varied between the two experiments.[3]

This seems to provide evidence against concepts comprising sets of necessary and sufficient conditions. There is both inter- and intra-subject disagreement concerning definitions of common nouns. If the classical view were correct, this should not be so—we should all, as competent language users, have invariant sets of conditions in our heads which we use at all times in defining concepts or in making decisions about category membership.

Although we appear unable to produce sets of conditions which are definitive of class membership, this does not in itself deal the death blow to the classical view. Perhaps our knowledge is not yet far enough advanced for us to produce the elusive defining conditions. Perhaps it never will be. The cumulative evidence against the classical view is fairly damning, however, and has led psychologists to look elsewhere for the roots of our concepts.

THE PROBABILISTIC VIEW

Struck by the apparent prevalence of typicality effects, Rosch and her colleagues proposed a new view of conceptual structure known as the probabilistic or family resemblance view.[4] The term "family resemblance" is not coincidental. Psychologists were becoming increasingly aware of the work of (the later) Wittgenstein, considering his notion of family resemblance a strategy with which it might be possible to bypass the problems plaguing the classical view.

Our concepts, according to the probabilistic view, comprise sets of characteristic or typical features which are abstracted across instances, rather than sets of necessary and sufficient conditions. These are mentally represented in the form of abstract summary representations. There need be no real-world instance corresponding to a representation, hence its abstract quality. Although many members of the category will possess all or many of the typical characteristics, there is no necessary requirement that they possess them all. Other

5: The Psychology of Categorization

members of the category will possess fewer of the characteristic features. We might, for example, say that characteristic features of the concept 'bird' are having wings and feathers, flying, singing and eating worms, and so these features will be included in the mental representation of the concept. Robins, thrushes and blackbirds possess all these features. Ostriches and turkeys, on the other hand, lack the last three features, yet they are still members of the 'bird' category. It is these overlapping chains of shared attributes which hold the members of a category together.

Swapping necessary and sufficient for characteristic features means that the probabilistic view can naturally incorporate typicality effects. Those category members possessing all or many of the characteristic features should be judged typical members, whereas those possessing fewer of the features should be judged less typical. In family resemblance terminology, this means that those members with higher family resemblance to the category (those possessing more of the characteristic features) will be judged more typical than those with lower family resemblance. We are therefore likely to view robins as more typical of the 'bird' category than ostriches.

The probabilistic view further assumes that the features associated with a category are weighted for saliency, which is calculated in terms of the number of category instances sharing a particular feature. Since all birds have wings, for example, wings are highly salient for the 'bird' category. An object's family resemblance to a category can thus be computed by summing all the weighted attributes that it possesses. A high sum will result in a high family resemblance score and vice versa for a low sum. Many birds, for example, have wings and feathers, fly and sing, so these features are highly salient for the 'bird' category and have high saliency weights attached to them. Robins possess all these attributes and so achieve a high family resemblance score, which explains why we consider them typical birds.

Rosch and Carolyn Mervis conducted a series of experiments to confirm the intuition that degree of family resemblance directly reflects typicality.[5] Subjects were asked to list attributes which they considered characteristic of each of 120 common objects. Each object belonged to one of six common categories and each of these items had been scored for typicality in its category in a previous experiment. Each attribute listed for each item falling into a particular category was weighted, according to how many items in the category had been listed by the subjects as possessing that attribute. Every item was then measured for family resemblance in its category by summing the weighted attributes which it had been listed as possessing. It was found that family resemblance scores correlated well with previous typicality ratings for the items. It seems that the more an item has attributes in common with other members of the category, the more it will be considered a typical member of that category. The probabilistic view, unlike the classical view, can account extremely well for typicality effects.

The probabilistic view can also accommodate the notion of fuzzy category boundaries. By definition, membership in a category is not an all-or-none affair, according to this view. Some objects may have a very low family resemblance score in a particular category, since they possess very few attributes in common with other members. It will then be unclear whether they belong to that category or to another, contrasting category. Is a rug I hang on the wall furniture, or is it a picture? Queries such as these are not anomalies, but an integral feature of probabilistic categories.[6]

Despite their differences, the classical and probabilistic views share a fundamental attachment to the notion of similarity. We group objects together because we see them as similar. Both camps construe similarity as something 'out there' in the real world, waiting to be perceived or discovered, something absolute and unchanging. Both camps consider similarity, in its own right, sufficiently constraining to provide an adequate explanation of the categorizations we make.

Yet, more recently, some psychologists have begun to doubt that similarity can do the job of explaining all our concepts and all the categorizations we make.[7] It is not clear that it makes sense to say that two things are similar *per se*. Similarity is not static in this way. Prior to judging two things similar, we must be aware of *what will count as similar in this context*. It becomes necessary to define a backdrop against which the notion of similarity can come into play. It is perhaps closer to the mark to say that we perceive two things as similar because we group them in the same category than to say that those two things are just similar and, by dint of their similarity, fall into the same category.

Gregory Murphy and Douglas Medin argue that, potentially, any two objects can be arbitrarily similar or dissimilar. They provide the example of a lawnmower and a plum which are similar in that they both weigh less than 10,000 kg, both did not exist 10 million years ago, both cannot hear, both can be dropped, both take up space and so on. We could equally make up a list of differences.[8] Similarity is too flexible to explain categorization. We need to drop down to a deeper level of explanation. We need to explain *why* we choose certain attributes over others on which to base our similarity judgements and to explain *in virtue of what* we judge members of a category to be similar. Why, for example, do we classify a Bedlington Terrier as a dog when it seems to share as many similarities with a lamb as with a Great Dane?[9]

In the light of this insight, asking subjects to list properties which they consider either defining or characteristic of particular concepts is explanation kicking in a stage too late. The attribute lists that those subjects provide will be biased in a fundamental way. Any attribute list will presuppose some sort of heuristic according to which those properties have already been judged relevant. It is the heuristic, then, which possesses the explanatory power, not the attributes in terms of which we formulate similarity judgements.

Lance Rips conducted an influential experiment illustrating that judgements of similarity may differ systematically from judgements of categorization.[10] He presented subjects with stories concerning members of both natural and artifactual kinds which underwent changes. Some of these changes Rips termed "essential" and others he termed "accidental."[11]

In one story, an animal was described as starting off life with characteristics typically associated with birds and as gradually maturing and taking on characteristics typically associated with insects. Rips considered this an essential change—one presumably determined by internal genetic structure. In another story, an animal exhibiting characteristics generally associated with birds was described as eating vegetation contaminated by hazardous waste. Following this, its behavioural and physical characteristics gradually became like those associated with insects. Rips considered this an accidental change, since the alteration involved surface characteristics only and was unlikely to alter the animal's genetic structure. 'Bird,' 'insect,' 'accidental change' and 'essential change' were not mentioned during the experiment, hence subjects had no idea of the ways in which they were expected to view the stories.

With artefacts, essential changes involved change in the designer's intended function and accidental changes involved changes in the artefacts' physical characteristics—such as color—which did not affect function. In one story, something fitting the description and function of an umbrella was redecorated to look more like something fitting the description of a lampshade, although it was still used to shelter its owner from the rain. In another story, subjects were told that an object fitting the description of an umbrella was designed with an intended function fitting that of a lampshade. Again, the words 'lampshade' and 'umbrella' were not mentioned during the experiment—subjects were simply given detailed descriptions of the imaginary objects.

The results show that in the case of both natural kinds and artefacts, accidental changes had a much greater effect on similarity judgements than on categorization judgements, while essential changes had a much greater effect on categorization judgements than on similarity judgements. Subjects tended to categorize the animal which underwent an accidental change as a bird, even though they judged it dissimilar to other birds and not at all typical of a bird. Conversely, the animal which underwent an essential change was, in its bird-like stage, considered very similar to and highly typical of birds, but was not considered likely to be a bird.[12] Likewise, the umbrella which underwent accidental changes was still categorized as an umbrella by subjects, despite bearing high similarity to lampshades. And the object intended for use as a lampshade by the designer was judged to be a lampshade, despite the fact that it bore high surface similarity to umbrellas. Subjects insisted that even if it were put to use in the future as an umbrella, it would still be a lampshade, since that was the designer's intended function.

Since similarity and categorization judgements can vary independently of one another, they are not inextricably linked, as the classical and probabilistic views assume. Rips' results suggest that, in the domain of categorization, there exist deeper explanatory principles which constrain, and can overrule, our notion of surface similarity. Similarity alone does not ground categorization.

The classical and probabilistic views share another assumption. They both treat our concepts as mentally represented in the form of features. Yet it is beginning to look less and less likely that our concepts are *only* represented by feature lists. Given our discussion of similarity, the question becomes why do we isolate the features that we do and what holds a list of features together? To say that a typical bird has wings and feathers, flies, sings and eats worms tells us very little—it is not a true representation of our knowledge, just an enumeration of properties. It omits our constant use of *relational* knowledge. We tend to assume, for example, that birds exhibit specific properties because they are related by some kind of underlying, genetic structure—it is this structure that brings coherence and life to the feature list. Perhaps, then, conceptual representation also involves some kind of relationship between features.

It seems likely that our conceptual knowledge also involves relations between concepts themselves. A concept does not possess life and meaning *per se*. The world is an essentially relational place and things tend to have relevance in relation to other things. A chair, for example, is made for *people* to *sit* on. Some chairs are used for sitting at *tables*. Others are made for *comfort* and we find them in people's *living rooms*, in front of the *television*, adorned with *cushions*. Or consider the concept 'dog.' Dogs are often *pets*, owned by *people*, they tend to live in *houses* and they like to eat *meat*. Our concept of the dog involves a lot more than a list of dog-like properties—it involves knowledge of how the dog operates in the world and how it is related to other things in that world. Concepts are not isolated entities. In order to grasp a concept, we require not only definitional, but also encyclopedic knowledge.

THE EXPLANATION-BASED VIEW

The inadequacies of the two similarity-based views of categorization have, in recent years, precipitated a move towards explanation-based theories of categorization. The key notion driving these approaches is that *our prior knowledge and theories about the world structure our concepts and provide unifying explanations for why we treat particular individuals as members of particular concepts or kinds.*

The approach as a whole remains in its infancy and the details are in the process of development, so it is not easy to glean a complete overview. Still, we can make some general remarks about the motivations and aims lying behind explanation-based theories. In particular, in response to those problems associated with similarity-based models, all versions of the explanation-based theory embrace the following three principles:

- Concepts are relational entities.
- Perceptual similarity does not prove sufficiently constraining to account for the categorizations we make.
- Concepts comprise more than feature lists and the categorization process amounts to more than straightforward attribute-matching.

First and foremost, then, the explanation-based view is a *relational* account. It is relational in two senses. First, it makes explicit the relations between properties associated with instances of a concept, construing these relations as causal or functional. Since all birds (natural entities) have wings, feathers and beaks, we tend to assume that these features must be *caused* by some kind of underlying genetic structure. This common genetic structure thus *explains* why birds fall together in the same class. Likewise, we assume that, since many chairs (artefacts) have both backs and seats, these features are necessary for chairs to *function* properly (for us to be able to sit on them). This common function *explains* why we group chairs in the same class.

Second, the explanation-based view makes explicit relations between different concepts. It is these relations or interactions that imbue those concepts with life and meaning. Unlike either the classical or probabilistic views, the explanation-based view acknowledges that in order to have a meaningful concept of anything at all, we must first understand the role that that particular kind of object plays in the world. We must appreciate its relationship to other kinds of thing and to ourselves. Our concept of the natural kind, gold, may include the knowledge that it is an element which is mined, that it is (or at least was, until very recently) stored in the vaults of the Bank of England, and that we tend to fashion wedding rings out of it. Gold becomes meaningful for us by dint of its relationship to many other concepts (elementary chemistry, miners and mining, the monetary system, matrimony) which play a part in our everyday lives. Similarly, our concept of an artefact such as a hammer may incorporate other everyday concepts—tools, toolboxes, picture hooks, pictures, nails, do-it-yourself projects and so on.

The explanation-based account is thus concerned with our knowledge of the world surrounding us. It recognizes that concepts are not isolated or self-contained. It recognizes that we play a part in shaping our concepts. Things are not categorized simply because of the way they are in themselves, but because of the role they play in our lives. Our concepts do not constitute straightforward mirror images of the world's categories—they are something much more cognitively interesting than that. In the words of William Wattenmaker, Glenn Nakamura and Douglas Medin, "Understanding human conceptual behaviour lies not with the nature of the world alone nor with the nature of human beings alone, but rather with the relationship between intelligent organisms and their environment."[13]

Although the explanation-based account does not do away with similarity completely, it takes much more of a back seat than in previous views. As we have already seen, functional or causal underlying relationships lead us to label things as similar and so to group them together. These underlying relationships serve to constrain, unite and impart meaning to our category groupings. They provide explanations for why we group particular individuals together. In some cases, straightforward perceptual similarity seems to work as an explanation. A thrush and a robin appear pretty similar—they both sing, fly, build nests, eat worms, and have wings, beaks and feathers—but what *really* explains our labeling them both 'birds' is our assumption that they both share in one underlying, probably genetic, structure. And we make the further assumption that this underlying structure causes their common perceptual and behavioral characteristics. But in other cases, straightforward perceptual similarity simply will not do. A whale surely looks much more like a fish than like a dog, horse or other mammal, yet still we—following the example of our biological taxonomists—class the whale as a mammal and not a fish. A deeper, underlying relationship—that of being warm-blooded and suckling young—grounds (or explains) our categorization and overrides surface or perceptual similarity and dissimilarity.

As we saw earlier in the chapter, any two objects can be similar to one another in an almost limitless number of ways, hence the question becomes why do we chose certain groups of similarity attributes over and above others for classification purposes? Why, for example, do we group robins with thrushes, swallows and pigeons, but apart from feather dusters, feather boas, and quill pens? The explanation-based view, as we have seen, provides an answer in terms of our theories and knowledge about the world. Robins, feather dusters, feather boas and quill pens are all similar in that they exhibit feathers, yet our knowledge of the world tells us that this kind of similarity is not relevant for categorization in this instance. Robins are living creatures, while all the other objects on the list are non-living artefacts. And the living/non-living distinction is one of the most basic we use when allocating individuals to kinds. Further, feathers are just one of a group of features that robins share with thrushes and swallows—a group of features that we assume arises from the birds' shared biological background, evolutionary history and genetic code. According to the explanation-based view, then, our concepts comprise not only lists of attributes, but also theories which explain why those particular attributes are relevant, which unite features within a concept, and which relate concepts themselves to the world as a whole. When we engage in categorization, we employ both world-theoretic knowledge and attribute-matching techniques.

To recap, then, the explanation-based view rejects surface, perceptual similarity as the ultimate basis of categorization. Instead, it proposes that our prior knowledge and theories about the world ground the categorizations that we make and so provide explanations for why we allocate particular individuals to particular kinds. Three further principles characterize the explanation-based view and distinguish it from similarity-based views of categorization—concepts are understood as relational entities; perceptual similarity is not considered sufficiently constraining to account for categorization; and concepts are said to comprise more than simple feature lists.[14]

Psychological Essentialism: An Explanation-Based Model

Psychological essentialism is a particular explanation-based model put forward by Medin and Andrew Ortony.[15] By exploring psychological essentialism, we will see how some of the general remarks about explanation-based theories cash out in terms of a concrete psychological model.[16]

The major tenet of psychological essentialism involves recognition of the psychological fact that our mental representations of things reflect our belief that those things have essences.[17] It is vital, however, to note two things. Firstly, Medin and Ortony are talking about our *representations* of things, not the things themselves (concepts are mental representations of entities). And secondly (as a consequence of the first point), they are not advocating metaphysical essentialism, which they dismiss as a "logically implausible doctrine."[18] Rather, their argument is that, irrespective of whether or not things have essences, we seem to *believe* that they do. They draw attention to the results of Rips's experiment which indicate that we believe entities possess defining essences which remain intact throughout surface changes. They also feel that we are likely to endorse essentialism, since most scientific enquiry aims to uncover the underlying nature of phenomena, rather than just to describe observable properties.

The problem with earlier accounts of categorization, explain Medin and Ortony, is that they deal only with surface properties of objects, which tend to be perceptual in nature. These are not adequate to explain how we categorize. Sometimes we group things together which are not perceptually similar, as in the whale-is-a-mammal-not-a-fish example. Furthermore, it is not clear that a family resemblance category coheres on the basis of surface similarity alone. A stool, a beanbag, a jumbo cushion and a swivel chair might all belong in the 'chair' category, yet it is not clear that they share sufficient perceptual similarity to form a cohesive group.

Medin and Ortony solve the problem by *linking* deeper and surface properties possessed by category members. There exists a strong, non-arbitrary connection between these two types of property, they claim, since the deeper ones constrain and, in some cases, generate the surface ones. Thus, we might assume that the 'real,' deep criteria for being an airplane (the ability to safely carry heavy loads for long distances through the skies) impose strong constraints on the surface features of airplanes, such as size of wings and shape of body. Or we might assume that, in the case of birds, the evolutionary role of flight and the birds' genetic structure generates certain surface features which include wings and hollow bones.

The attributes associated with a concept lie on a continuum of accessibility, Medin and Ortony suggest, stretching from highly accessible to relatively inaccessible and hidden. At the very deepest level, these attributes fill an "essence placeholder" in our conceptual representations, which may be filled with different kinds of thing, depending on the concept in question. In some cases, this may be a list of defining conditions. We might, for example, consider that being an element with atomic number 79 comprises the essence of gold. In other cases, this may be a complex set of beliefs more akin to a theory than an attribute list. Gregory Murphy and Medin provide the example of categorizing someone at a party who jumps into a swimming pool fully clothed as intoxicated, even though we have never seen a drunk person behave in this exact way before. In this kind of situation, we engage in an inference process, they reason. We have a general theory of how drunks behave (inappropriately) and of the type of situation in which people drink too much (parties) and, by using this theory, we conclude that the most likely explanation for that person's behavior is that he/she is drunk.[20] The essence placeholder might also contain the belief that there are experts who can identify the essence of the thing, even if laypeople cannot. The beliefs filling this placeholder constitute the essence of psychological essentialism and so constrain surface properties.

It is this constraint of surface by deep properties that allows us, on an everyday basis, to successfully resort to surface perceptual similarity for classification purposes, Medin and Ortony argue. In general, we can rely on the heuristic that appearances do not deceive, but occasionally—as with the whale example—this heuristic breaks down. And, as is so often the case, the exception reveals to us the mechanics of the rule. Our notion that entities possess essences that constrain their surface appearance functions as an explanation of how the world works. This theory of essence allows us to categorize and to make predictions and inferences about the entities populating the world. It also explains why we make the categorization judgements that we do.

And so we see the distinction between earlier and later theories of categorization. A shift clearly occurs from a very tightly defined notion of what concepts are and what categorization involves to a much more flexible, less formal approach. A clear shift also occurs from the notion of humans passively reflecting real-world categories through their concepts to the idea that humans make an active contribution to their concepts and to real-world categories.

Categorization is a very simple and precise affair, according to the earliest classical view. Our concepts become no more than simple mirror images of the categories assumed to exist in the world. Every natural category possesses a set of invariant, objective conditions which determine membership in that category. All *we* have to do is

5: The Psychology of Categorization

discover them. We make no active contribution to categorization. Our role is a passive, receptive one. The classical view draws a harsh line between humans and the environment in which they operate. We stand apart from the world, revealing and reflecting that world as it is, in itself.

Although the probabilistic view begins to recognize some of the faults of the classical view, still the changes it introduces concern *conceptual structure only*. Its proponents may have realized that an all-or-none defining structure cannot apply to many of our concepts, but they paid no real attention to our role in the environment. Still humans are viewed as passive mirrors of real-world categories. Although family resemblance theorists note that certain characteristics frequently co-occur in the world (wings, beak and feathers usually co-occur, for example), they make no attempt to explain *how* we relate such features together when categorizing.

Explanation-based theories adopt a very different approach to categorization. Finally it is acknowledged that our concepts may constitute more than passive reflections of pre-existing categories. Finally it is acknowledged that we play some part in shaping both our concepts and real-world categories. We actively contribute to the categorization process by constructing theories or explanations which underlie our similarity and categorization judgements. As a result of our interaction with the world, we relate particular classes to other classes and so those relations which are pertinent *for us* shape our conceptual representation. The encyclopedic knowledge we use when categorizing does not simply constitute a reflection of the one way in which the world is, but involves us in theory-construction and in adopting specific perspectives on the world. The environment is no longer self-contained and the division between human agent and world begins to blur. As Medin puts it:

> It is tempting to think of categories as existing in the world and of concepts as corresponding to mental representations of them, but this analysis is misleading. It is misleading because concepts need not have real-world counterparts (e.g. unicorns) and because people may impose rather than discover structure in the world. I believe that questions about the nature of categories may be psychological questions as much as metaphysical questions. . . . Part of the answer to the categorization question likely does depend on the nature of the world, but part also surely depends on the nature of the organism and its goals. Dolphins have no use for psychodiagnostic categories.[21]

6

Philosophy and the Psychology of Categorization

Despite a common subject matter, the psychology of categorization has gone almost unnoticed by philosophers interested in classification. One philosopher—Georges Rey—does, however, discuss some of the psychological work, although his aim is destructive rather than constructive. He focuses an attack on Edward Smith and Douglas Medin's classic review, *Categories and Concepts,* which covers the classical, probabilistic and exemplar views, but predates the explanation-based view. Smith and Medin's major conclusion therein is that the findings of psychological studies of categorization do not fit the classical view of concepts.

As the discussion progresses, we will see that Rey is an objectivist philosopher *par excellence,* believing that the classical view of categorization is correct, that classification concerns metaphysics alone, and that psychology and epistemology tell us nothing about the nature of concepts. We will also see that, at this stage, the psychologists agree that their work deals in epistemology alone and so has no relevance for metaphysics.

Much of the debate centers on Smith and Medin's core and identification procedure—a psychological theory of categorization which they propound in *Categories and Concepts.*[1] This theory distinguishes between a concept's core (which contains deep diagnostic properties) and the procedure we use to identify that concept (the properties we commonly use to categorize its real-world instances). When making decisions about concept membership, we rely mainly on identification properties—properties which tend to be perceptual and easily accessible, but not always foolproof. Core properties are much more reliable, but less accessible, and so less available to us when making decisions about concept membership.

6: Philosophy and the Psychology of Categorization

Consider gender. The identification properties in this case may include things like clothing, hairstyle and voice, while core properties might involve possession of sexual organs. Or consider gold. Identification properties might include its color, the uses to which it is put and where it is mined, while core properties might involve being a basic element with atomic number 79.

The core and identification procedure aims to combine the classical view with the typicality effects associated with the probabilistic view. A concept's core might have a classical-type structure—perhaps being composed of essential, defining conditions—while the identification procedure may have a probabilistic structure—identification properties may simply be characteristic of the class, those members considered typical of the class possessing a larger number of the characteristic properties. We can thus view the core properties as a means of backing up or justifying 'quick and dirty' judgements made on the basis of identification properties alone. We might, for example, come across a substance looking very like gold, but on closer inspection find that it is not gold, since it is not a basic element with atomic number 79—it does not possess the requisite core properties.

It is vital to bear in mind here, as with the model of psychological essentialism discussed in the last chapter, that psychologists deal with epistemology only. The distinction between the core and identification procedure does not, therefore, mark a distinction between metaphysics and epistemology, rather the core properties are simply less accessible but more reliable than identification properties for making epistemological decisions about concept membership. And this is so irrespective of whether there are such things as metaphysical core properties. Smith, Medin and Lance Rips stress that by distinguishing core properties they "are trying to characterise the layperson's theory of the nature of things . . . [not] to characterise the nature of things."[2]

The goals driving Rey's attack on Smith and Medin are two-fold: (a) he wants to question that *psychological* findings have any relevance for a theory of concepts and (b) he wants to shore up the classical view of concepts.

He begins by driving home his belief that the psychology of categorization continuously confuses questions of metaphysics with questions of epistemology. More precisely, he thinks it confuses questions of conceptual *identity* with ones of conceptual *access*. Psychology may provide interesting results concerning conceptual access, but this tells us nothing about concepts themselves, or their identity conditions, or what conditions must hold in order for us to be able to say that someone *has* a particular concept.

Rey draws a sharp line between metaphysics and epistemology or between "issues surrounding *how the world is* (what exists, what is true) and issues surrounding *how we know, believe, infer how the world is*,"[3] arguing that concepts can play a metaphysical role by providing the basis for certain metaphysical claims. The concept 'cow' can provide the basis for the claim that Elsie is a cow by specifying certain facts about Elsie in virtue of which she is a cow. "Concepts in this role may be regarded *in isolation*, as providing the principles of classification for what is to count as their instances," he concludes.[4] Concepts also play an epistemological role, which involves our method of ascertaining whether something is an instance of a concept. There are, for example, metaphysical conditions in virtue of which someone is either male or female, but in order to ascertain someone's gender, we rely on extraneous features such as hairstyle, mode of dress and pitch of voice. These are not features in virtue of which someone is either male or female, but perceptually salient features which we use to arrive at reasonable decisions concerning category membership. The psychology of categorization tells us much about the latter epistemological role of concepts, Rey claims, but nothing at all about what he dubs the "only serious notion" of concepts—their metaphysical role.

Rey insists that the word 'categorizing' can refer either to how things are *correctly* categorized (to metaphysics) or to how people engage in the process of categorizing (to epistemology), suggesting that Smith and Medin themselves consider this distinction when they differentiate between a concept's core and its identification procedure. But he concludes that they fail to take this distinction seriously and see no reason to differentiate between the metaphysical and epistemological roles that concepts play. Rey, however, insists that this is a vital psychological distinction. If we fail to make this kind of distinction, we will have to say that different people who use different identification procedures for the same concept, or one individual who uses different identification procedures for the same concept at different times, have different concepts. Suppose two people use slightly different perceptual features for deciding that a robin is a bird—surely we would not want to say that these two individuals have different 'bird' concepts. Furthermore, if no distinction

6: Philosophy and the Psychology of Categorization 69

is made, any change in belief about how to tell whether or not something is an instance of a concept will have to count as a change in concept.

Rey is concerned that the epistemological role of concepts—and so the discipline of psychology—can account for neither conceptual identity nor conceptual stability. Only concepts in their metaphysical role can account for these phenomena, by providing a list of defining conditions in virtue of which something is the thing that it is. What makes two things instances of the same concept is that they conform to the same definition. What constitutes people having the same concept is that the concept has one and the same definition attached to it, despite the fact that different people may associate different identification procedures with it.

Rey's second goal—defense of the classical view of concepts—turns on what he calls "an entirely implicit assumption attached by Smith and Medin to the Classical View."[5] This is the assumption that in order to have a particular concept, one must *know* its defining conditions. Rey, however, believes that whether anyone knows that a concept has defining conditions remains entirely irrelevant to whether or not it *does* have defining conditions. He posits the Hypothesis of External Definitions, which asserts that "the correct definition of a concept is provided by the optimal account of it, which need not be known by the concept's competent users."[6] Thus, although there probably are scientific experts who can provide definitions (optimal accounts) for many of our ordinary language terms, there need not be. Definitions are divorced entirely from anyone's ability to provide them. There will always exist conditions in virtue of which something is the thing that it is, quite independent of anyone's knowledge of those conditions.

The Hypothesis of External Definitions effectively protects the classical view from all the psychological evidence adduced against it, Rey believes. By maintaining an unerring division between metaphysics and epistemology, what people know or believe remains irrelevant to the real structure and true classes of the world—those characteristics that concepts should reflect. Since psychology, in Rey's eyes, talks only to epistemology, its findings leave metaphysical issues untouched and a haven remains in which the classical view is able to survive, fit and well.

Rey is, of course, an out-and-out objectivist, in the tradition of (the early) Putnam and Kripke. He supports objectivist metaphysics by indicating that the world possesses its own unique category structure as determined by essential properties, a structure quite unaffected by human cognitive activity. He also supports objectivist epistemology with his talk of correct categorization involving reflection of the world's inherent category structure. By making it clear that the definitions which attach to metaphysical categories are simply a matter of the way the world is and have nothing to do with our knowledge or understanding of either those definitions or the world in general, Rey confirms the partition between metaphysics and epistemology. He wipes psychology off the metaphysical agenda for good.

Smith and Medin (together with Lance Rips) remain unperturbed by the ferocity of Rey's attack. They claim total agreement with Rey's argument that in order to provide identity conditions for concepts, we must concern ourselves with metaphysics. They concur that different people possess the same concepts (irrespective of identification procedure) and that, while concepts remain constant, properties used to identify concepts may not. They have good reason for remaining unperturbed, of course, since Rey has misunderstood their distinction between the core and identification procedure. He assumes that a concept's core involves metaphysics and that its identification procedure involves epistemology, yet Smith and Medin quite explicitly stated that neither has anything to do with metaphysics—both deal with epistemology. They therefore never claimed to deal with anything other than the epistemology of categorization, contrary to Rey's belief that they drew a distinction between metaphysics and epistemology and then chose to ignore it.

What, then, is the point of difference between Rey and the psychologists? Smith, Medin and Rips suggest that Rey, in stressing the need for metaphysically defining conditions, is attempting to characterize the nature of things—a notion involving scientific experts who (possibly) can specify what those defining conditions are. In contrast, they (as psychologists) are, by talking about core properties, attempting to characterize the layperson's theory of the nature of things—which involves only mental representation and epistemology.

What of Rey's claim that epistemological categorization cannot account for either inter- or intrapersonal conceptual stability? Once again, Smith, Medin and Rips agree with Rey that if we equate stability of concepts with sameness of concepts, then epistemology will not be able to account for stability, since same identity conditions in-

6: Philosophy and the Psychology of Categorization   71

volve metaphysics. Yet epistemology can still play a role in conceptual stability, they insist.

There exists another kind of stability—communality—which equates with similarity of mental content (epistemology). When two people exhibit similar mental content, we may conclude that they share the same concept.[7] It is also likely, the psychologists argue, that core properties provide another source of communality. We all tend to share basic core beliefs, such as that instances of a particular animal share a common genetic structure, or that offspring must be of the same biological type as their parents. Interchange with others enhances this interpersonal communality, since we tend to point out one another's misclassifications and alter our use of identification and core properties accordingly. Communality is, for the purposes of everyday comprehension and communication, quite adequate to secure conceptual stability without invoking sameness of concepts and so metaphysics, Smith, Medin and Rips conclude.

This debate between Rey and the psychologists provides a convenient entrée into one of the issues with which my account of scientific classification deals. I wish to challenge both the unfaltering, absolute line that Rey (and other objectivists) draw between metaphysics and epistemology and the superior, determining role that they accord to metaphysics. Rey suggests that, if anyone can uncover the defining conditions or true nature of things, it will be our scientists, since they study metaphysics, pure and simple. I instead suggest that *even in the case of expert science*, classification involves a synthesis of metaphysics and epistemology. As internal realism shows, we simply cannot divorce the way the world is from our means of accessing the way the world is.

At the time of this debate, the psychologists themselves, when challenged, concurred in the drawing of a line between metaphysics and epistemology, placing themselves, as psychologists, squarely in the domain of epistemology. At every opportunity, they insulate psychology from metaphysics, rather than suggest that psychology may inform metaphysics. They thus agree that only metaphysics can provide identity conditions and sameness of concepts and, in order to save some role for epistemology in conceptual stability, they introduce communality and similarity of mental representation. They appear, like Rey, to wholeheartedly accept that a predetermined, fixed metaphysics provides the key to *correct* classification of objects in the world.

Despite this overt agreement, Smith, Medin and Rips's theorizing almost seems to contradict their words. By distinguishing between core and identification properties and by claiming that the former back up the latter, it is as if they are claiming that it is not simply the way the world is, but rather the deeper theories we have about the world that ground our categorizations. In other words, they might almost be claiming that epistemology *does* have relevance for metaphysics, if it were not for their emphatic insistence that neither identification nor core properties have anything to do with metaphysics.

Yet this possibility fades when Smith, Medin and Rips claim that Rey is trying to characterize the nature of things and so speaks in terms of the special sciences, while they are trying to characterize the layperson's theory of the nature of things and so speak in terms of lay science. This suggests they too believe that things in the world possess a predetermined nature and that the role of the special sciences is to discover the metaphysically true facts in virtue of which things are what they are. They seem to believe that only non-experts have theories about the way the world is. Expert scientists work differently. They do not deal in theories, but search for a correct classification of the objects in the world, which involves reflecting an unbiased, context-free picture of the world's inherent structure.

But do expert scientists really work without the aid of unifying theories? Are scientific classifications really devoid of all epistemology? Does 'correct' classification involve nothing more than reflection of metaphysical structure? In the last chapter, we saw that, with the advent of explanation-based accounts of categorization, there has been a blurring of the line between metaphysics and epistemology. Medin suggested, for instance, that "people may impose rather than discover structure in the world" and that "questions about the nature of categories may be psychological questions as much as metaphysical questions."[8] It seems as if psychologists like Medin are at last becoming aware of the role that epistemology plays and are beginning to realize that perhaps psychology has something to offer metaphysics after all.

In the next chapter, I extend the insight that epistemology has relevance for metaphysics to scientific classification. We will see, among other things, that the explanation-based account of categorization can apply not only to laypeople, but also to expert scientists.

7

Five Interrelated Theses: The Theory Explained

I want now to present my own pluralist theory of scientific classification. I offer this as an alternative to the objectivist interpretation of scientific classification, as outlined in chapter 2. In particular, I offer it as an alternative to Objectivism's assumption that the world divides uniquely and inherently into a number of natural kinds, delineated by essences or sets of defining conditions, and that the role of science is to uncover and reflect that inherent taxonomy.

Over the preceding chapters, a background has emerged against which a pluralist account of scientific classification is plausible. We have come to understand the failings of Objectivism. We have seen how internal realism, plus a contextual correspondence theory of truth, acknowledge the active role that human agents play in their cognition of the environment, while preserving our common-sense understanding of realism. We have learned that the classical view of categorization fails to account for the structure of laypeople's concepts. We have suggested that classification may be as much a matter of epistemology as of metaphysics, not only among laypeople, but also among scientific experts—the very people who, Objectivism insists, impartially adopt a God's Eye view of reality.

Informed by this context, I suggest that the following five interrelated theses characterize scientific classification:

- Scientific classification is not clear-cut; rather it is an inherently messy affair.
- Scientific classification involves more than the way the world is (or metaphysics).
- Scientific classification involves human input in the form of theories, explanations, aims and commitments (or epistemology).

- Scientific classification therefore comprises a synthesis of metaphysics and epistemology.
- The explanation-based model of categorization applies to expert scientific classification.

At the base of my account lies a commitment to a world outside ourselves, a world which comprises entities and exhibits characteristics quite independent of human existence, thought, or action. I consider this world to be an extremely rich and complex place, so much so that it has numerous patterns of similarity (or regularity) and difference running through it, patterns which criss-cross over the entities that populate that world. The richness of this environment means that, contra Objectivism, the world does not come with a preferred description—it does not come with a preordained set of natural kinds and attached essences. In this case, scientific classification cannot be a matter of reflecting the world's inherent structure, since no such unique structure exists. Rather, scientists are faced with a choice regarding which of the many regularities or patterns of similarity are to count as relevant for classification purposes. In order to get a classification up and running, scientists must therefore adopt a theoretical or explanatory perspective—an heuristic which makes certain patterns of similarity or regularity stand out as salient for classificatory purposes.

Since different scientists may adopt different heuristics, controversy can spring up regarding which theoretical perspective ought to be adopted and competing classifications can arise regarding either the defining properties of a particular class, or the entities to be included in a particular class, or both. The inherent complexity of the world means that it can support competing groupings of the entities within it, hence classification can never be the straightforward, cut-and-dried affair of which the objectivist dreams.

In order to be an adequate, successful competitor, a classification must be based on real and objective properties shared by entities. (If, for instance, scientists take atomic number as the principle of classification of the chemical elements, they must abide by the fact that each chemical element has a certain number of protons in the nucleus of the atom—a fixed atomic number. They are bound by the fact that gold, say, has 79 protons in the nucleus of the atom and not 78.) It is *because* competing classifications reflect different, yet real, aspects of the natural world that the world can neutrally support these competitors, rather than preferring one over the others. Since the world is neutral in this way, classification must involve more than a reflection of the way the world is. It must involve more than pure metaphysics. Since the world 'is' in more than one way, some form of human input is required to select one or more of these ways, converting it/them into the basis of a (system of) classification. Scientific classification therefore involves a synthesis of metaphysics (one of the ways in which the world is) and epistemology (theories, explanations, aims or commitments which allow the scientist to isolate one of the different ways in which the world is).

All these points combine, I suggest, leading to the conclusion that the explanation-based model applies not only to lay classification, but also to classification conducted by expert scientists. Since, on the pluralist account, entities can share a number of different similarities, an underlying theory or explanation is required to account for why certain attributes or similarities are chosen over and above others as salient for categorization. The reliance on an underlying theory or explanation implies that scientific classification involves more than straightforward attribute-matching. It also implies that scientific concepts are relational entities, involving wide, encyclopedic understanding, rather than isolated entities characterized in terms of lists of defining, or necessary and sufficient, conditions.

I am going to discuss these five interrelated theses with reference to three case studies taken from the history and sociology of science, each of which covers a controversy between scientists over classification. The fact that such controversies have existed and, in some cases, continue to exist, speaks strongly in favor of a pluralist account of scientific classification. With the aid of these studies, we can see how the pluralist model works in concrete instances and how it accounts for complexity and dispute within taxonomy.

JOHN DEAN'S "CONTROVERSY OVER CLASSIFICATION"

John Dean catalogs the controversies surrounding two very different methods of botanical classification—orthodox (Linnean) taxonomy and more modern experimental taxonomy, or biosystematics. This provides a particularly interesting example, since a similar, unresolved controversy continues to the present day.[1]

Orthodox taxonomy holds that classification must depend on

morphological discontinuities between plants which are capable of discernment through perception alone, rather than through experiment. Morphological difference is taken, by traditional taxonomists, as the one objectively real, metaphysical property that grounds plant classification. Why? These traditionalists believe in the essentialist precept that species are entities which exist objectively in the natural world and are individuated by possession of real essences or characteristics. It is only by concentrating on morphological discontinuity, they believe, that is it possible to arrive at (or discover) a classification which reflects this natural structure and built-in division. The orthodox attachment to essentialism acts as the underlying theory or explanation which highlights morphological discontinuity as salient for matters of classification. Orthodox taxonomists bring this prior epistemological commitment to their work and, as a result, concentrate on variation between species to the exclusion of other distinguishing properties.

Orthodox taxonomy is also eminently *practical*. It enables its users to classify plants with ease and in comfort, by applying standardized descriptions and precise rules in their examination of dead plant material in the herbarium. It allows accurate storage and rapid retrieval of information, making it particularly easy to fit new plant species into the confines of the established system. The desire for convenience and precision in classification also acts as a relevant explanatory factor, combining with the belief in essentialism to prompt orthodox taxonomists to classify on the basis of morphological discontinuity.

Criticism of Linnean taxonomy first arose with the development of new theories and practices in fields outside taxonomy. Following the rise in knowledge about, and interest in, genetics in the years following 1900, the new fields of (cyto)genetics and cytology (the study of cells) came into being. These introduced a novel set of techniques—such as transplanting and crossability experiments and detailed chromosome analysis—which paved the way for an alternative taxonomy.

The new experimental practices suggested to some botanists that orthodox taxonomy was inadequate for describing intraspecific variation. These biosystematists instead argued that classification should be based on another real, metaphysical property shared by specific plants—the ability to interbreed and exchange genes. Behind their choice of this property lay an epistemological commitment to the importance of experimental methods and the categories that such methods produce. Crossability—which investigates the limits of gene-exchange between populations—was the most important experimental method and so constituted the underlying theory or explanation in terms of which facts about interbreeding and gene exchange became salient for classification.

Interestingly, Dean tells us that orthodox taxonomists agreed with biosystematists that evolutionary aspects of a plant's history are important. Yet they denied, on the grounds of practicality, that these aspects make a good basis for taxonomy, since classifications produced using experimental methods are slow and require prolonged research. We may conclude that, while traditionalists recognized the relevance and reality of experimental methods and properties *in general*, still their own prior epistemological commitment to speed and ease in classification prevented them from considering interbreeding and gene exchange salient for *taxonomic purposes*.

The botanical dispute did not end with competing underlying theories—these two systems actually produce competing classifications, particularly at the species level. Dean provides the example of the plant genus Gilia, which contains two species-complexes—Gilia inconspicua and Gilia tenuiflora. The Gilia inconspicua complex comprises five interrelated sibling species which, although fertile in themselves, are all highly intersterile. From the point of view of the biosystematist, all five form separate species. Morphologically, however, all five form a single group, since it is only through microscopic examination of chromosomes that they can be distinguished. Similarly, the Gilia tenuiflora-latiflora complex comprises at least four different elements which are morphologically distinct, yet capable of gene exchange. The orthodox taxonomist claims at least four separate species here, while the biosystematist claims only one.

In each case, classification is based on real properties—either morphological similarity/discontinuity or the ability to interbreed and exchange genes. We cannot, therefore, judge one decision right and the other wrong. Each classification is correct or true when considered along with its theoretical context. Given an orthodox background, the Gilia inconspicua complex forms one species—its members really do form a morphologically continuous group. Given an experimental background, the same complex comprises five separate species—all five elements really are intersterile. Each system makes observations which correspond to aspects of the real world—the world supports each equally. Each classification is therefore true, so long as we understand truth as correspondence, given a particular context. It is only when we synthesize this context (epistemological stance) with the real properties that the context highlights (metaphysical aspects) that we gain a complete picture of the classificatory process. It then becomes clear how categorization amounts to more than straightforward attribute-matching and instead presupposes a wider theoretical context.

Adrian J. Desmond's "Designing the Dinosaur"

This case study differs from the last in that it charts the historical 'creation' of a category (the dinosaur) by one Richard Owen.

Prior to Owen, there was a healthy paradigm in place which considered Owen's dinosaurs to have been enormous reptiles—"fossil lizards." This paradigm relied on metaphysical properties common to the jaws and teeth of both the fossil dinosaurs and extant lizards. But, in his "Report on British Fossil Reptiles" made to the British Association in 1841, Owen placed these creatures in their own taxonomic category—*Dinosauria*. His decision relied on a different set of metaphysical properties—anatomical peculiarities of the sacrum, ribs and extremities, together with enormous size—which this time served to distinguish the dinosaurs from living lizards and Mesozoic marine saurians (lizards), but to liken them to large mammals. Owen went further, using these anatomical references to make bold conjectures about the vital ecological significance of dinosaurs. He also insisted that, although cold-blooded, these creatures were akin to pachydermal mammals, claiming that since they had the same thoracic structure as crocodiles, they must have possessed a four-chambered heart and that since they were so well adapted to life on earth, they must have had a circulatory system close to that currently possessed by warm-blooded vertebrates. Yet these animals for which Owen made so many hypotheses were only scantily known on the basis of three species.

Owen introduced a new paradigm for estimating dinosaur size, which involved measuring individual fossil vertebrae, then estimating their total number using existing crocodiles and lizards as prototypes. Yet he was interested in more than measurement here, managing simultaneously to effect a gross morphological transformation of the animals. He replaced the paradigm of small, lizard-type legs

7: Five Interrelated Theses

with mammal-type legs and claimed that, due to their massive weight, his dinosaurs would have stood upright on all fours, rather than sprawling close to the ground. This was again a puzzling conjecture, given that Owen had so little evidence to support his position.

Why was Owen so keen to claim that his dinosaurs were more akin to mammals than reptiles, in opposition to his predecessors and on the basis of scant evidence? What kind of theory, explanation, or epistemological stance highlighted, for Owen, their mammalian properties to the exclusion of their reptilian similarities? Desmond tells us that, "Whereas Cuvier, Buckland, Mantell, and von Meyer all noted isolated anatomical similarities between "Fossil Lizard" and mammal bones, they persisted in using the lizard blueprint for reconstruction. . . . Unlike his predecessors, Owen *used* this mammal quality."[2] Clearly, Owen's predecessors recognized that dinosaurs shared properties in common with mammals. These similarities existed independently and objectively in the real world for all to see. So what motivated Owen to isolate and work with these properties? And why did his predecessors concentrate on the alternative set of reptilian similarities?

Owen's aims were two-fold, Desmond suggests. Firstly, Owen wanted to defeat Lamarckism (also known as transmutation in this context), which posited a huge sequence of life forms extending from the simplest to the most complex, with human beings at the pinnacle of perfection. Animal organs were possessed of some kind of internal energy or excitation, Lamarck argued, which impelled these organs to become more complex, with the result that life forms gradually took their place on successively higher levels. Secondly, Owen hated and so wanted to destroy one of Lamarck's eminent followers, Robert Edmond Grant.[3] Grant believed that evidence of Lamarck's system would be imprinted on the fossil remains, with the earliest fossils of forms at their most primitive and later fossils showing a gradual rise in the complexity of these forms. He therefore aimed to provide an historical construction of Lamarckism.

By exploiting the dinosaurs' mammalian similarities, Desmond suggests that Owen was able to fulfill both his aims simultaneously. Owen's reconstruction showed that reptiles at their peak (as dinosaurs) approached the complexity of the mammal form both morphologically and physiologically. Yet this occurred way back in the Mesozoic era. If this was so, reptiles must have later suffered a degeneration (something not allowed by Lamarckism), resulting in the contemporary large numbers of small lizards. Owen's mammalian dinosaurs combined with other evidence[4] to disprove the transmutation argument for a continual increase in fossil complexity through time. Lamarckism fell from grace and so, too, did its prominent supporter, Grant.

We can see that Owen's strong intellectual dissatisfaction with Lamarckism acted as the underlying explanation which made the dinosaurs' mammalian qualities salient in his eyes for classification. By focussing his attention on this set of similarities, by rejecting the lizard blueprint in favour of a mammal blueprint, he was able to elevate the dinosaurs to their own taxonomic rank, so displaying the inconsistencies of, and ultimately overthrowing, Lamarckism. Owen's hatred for Grant added fuel to the fire. It provided a second, underlying aim or motive which combined with his abhorrence of Lamarckism to push Owen into exploiting these mammalian properties.

Owen had aims and commitments which his predecessors lacked. Although Desmond does not spell out what motivated Owen's predecessors to exploit the dinosaurs' lizard-like properties, still we can speculate. We know that they were happy to continue with the established paradigm, despite their recognition of the mammalian properties. Why? They, unlike Owen, had no quarrel with either transmutation theories or with Grant. They had no reason for showing that these animals were far more advanced than currently existing lizards. And so they stuck by the familiar, workable paradigm. Only through sharing in Owen's epistemological stance would the mammalian similarities have become sufficiently salient for the earlier scholars to have used them in making classificatory judgements. It is the *synthesis* of metaphysical and epistemological aspects that produces classificatory decisions. Metaphysical aspects alone lack the necessary prominence.

Owen proceeded to use his dinosaur evidence in arguing for degeneration over transmutation. The contemporary climate suited his purposes. Not only did the evidence tell against transmutation theories, but they were considered anti-Christian, the notion of each life form having *within itself* the power of self-improvement taken to imply materialism. And so we have yet another underlying explanatory factor which combined with Owen's hatred of Lamarckism in general and of Grant in particular, making the dinosaurs' mammalian properties salient for classification in Owen's eyes. He saw his opportunity and, already sensing success, forged ahead to exploit and promote the mammalian quality to the exclusion of the reptilian similarity.

In response to the fossil evidence, a compromise position called Discontinuous Progression was eventually reached. Life did advance, it was argued, but step-by-step rather than continuously. Each class was said to appear at a discontinuity and, once created, to stay at that level or even to decline. Since there was no continuous link between classes, it was claimed that God must have intervened on each occasion to create each new class. Discontinuous Progression therefore preserved succession of classes, degeneration and God's omnipotence. Transmutation theories, by contrast, failed to account for degeneration and threatened the supremacy of Christianity. They fell by the wayside and the lizard paradigm fell along with them. Nevertheless, we should be aware that circumstances—and not reality—conspired to discredit the lizard paradigm. Although the paradigm fell, the similarity between lizards and dinosaurs still existed. The change in epistemological stance altered—Discontinuous Progression replaced transmutation—and so the lizard-like quality was eclipsed by the new salience of the mammalian similarity. Categorization, then, is not simply a matter of attribute-matching. It is only in the wider context of the clash between transmutation and degeneration that we can understand how different scientists were able to exploit different types of similarity or attribute and to understand how this led to a reclassification of the dinosaurs.

Nor should we assume that Owen's findings ended controversy over dinosaur classification. These animals are currently considered to have been members of a (now extinct) race of saurian (resembling lizards) reptiles. The fossil remains point to organisms which in some respects resembled birds and in other respects resembled mammals, and here the debate continues to rage. In January 1999, scientists completed study on the best-preserved dinosaur yet found—an infant Scipionyx samniticus, a theropod or meat-eater. They argue that its lungs and other structures bear little similarity to modern birds—it exhibits body cavity partitioning only seen in living animals that use an active diaphragm to help ventilate their lungs, like mammals and crocodilians. By contrast, those scientists in favour of the theory that dinosaurs were the ancestors of modern birds[5] point to the recent discovery of a Caudipteryx zoui, which exhibits the features of a theropod dinosaur and had feathers covering its body. We cannot say that one scientific camp is right and the other wrong—some dinosaurs really do bear similarity to modern birds (feathers, in this example), while other dinosaurs bear more similarity to mammals (body cavity partitioning, in this example). Perhaps the debate will continue to rage, unresolved, or perhaps one camp will gain ascendancy over the other, but still we should remember that both sets of similarities are equally real. We will just have to wait and see whether (and how) one epistemological stance elevates the salience of its associated set of attributes to the exclusion of the other.

Mary P. Winsor's "Barnacle Larvae in the Nineteenth Century"

Mary Winsor's case study deals with the finding by field naturalist John Vaughan Thompson that barnacles begin their lives as larvae and indicates "the active role theoretical considerations play in the process of zoological classification."[6]

By 1823, Thompson had begun to suspect that many plankton were none other than the young of species already known to zoologists. In 1826, he captured some "bivalved crustacea with stalked eyes," managing to keep them alive long enough that they moulted and became barnacles.

This was an important finding for two reasons. Firstly, it had previously been assumed that one of the factors separating the higher from the lower crustacea was that the lower underwent metamorphosis, while the higher did not. Absence of metamorphosis was also one characteristic sometimes used by taxonomists to separate the higher crustacea from insects. Secondly, it had previously been thought that barnacles did not belong to the class of crustacea at all. Linnaeus grouped them with mollusks on the basis of their calcareous shells and lack of segmentation and, in 1809, Lamarck claimed that they formed a transitional class (Cirripedia) between the annelids (worms with segmented bodies) and the mollusks. As a result of his discovery, Thompson was convinced that barnacles were neither mollusks nor cirripedes, but belonged with the crustacea.

Prior to Thompson, no one had considered that barnacles might be members of the Crustacea class, since no one knew that they began their lives as larvae and underwent metamorphosis. Classification of barnacles had been determined purely on the basis of the adult form, exploiting those metaphysical properties—the calcareous shell, evidence of hermaphroditism and the visceral anatomy (the large internal organs of the body)—remaining after the moult. Yet Thompson was prepared to concentrate on the metaphysical properties exhibited by the larval form—moulting, jointed limbs and crustacean-like jaws—to the exclusion of those exhibited by the adult, leading to his reclassification of barnacles as crustacea. What underlying theory or explanation prompted Thompson to focus on larval characteristics to the exclusion of other distinguishing features?

In 1827, Thompson caught some female crabs along with their eggs (on the abdomen). What emerged from the eggs were not baby crabs, but zoeae. On the strength of this further evidence, he published his results, claiming that metamorphosis such as occurred with the common crab was a general principle among the higher crustacea. His research was repeated and accepted over the next few years.

In 1830, Thompson published his material about barnacles, describing the metamorphosis of his stalk-eyed crustacea into sessile (non-stalked) barnacles and hypothesizing that stalked barnacles would have similar larvae to their sessile cousins. Later that year he was able to test this hypothesis when ships came in to Cork Harbour with stalked barnacles plus ova on their bases. The larvae that hatched, however, were quite different from those of the sessile barnacles. He therefore concluded that the two types were less closely related than he had supposed.[7] "This was a reversal of opinion from his earlier memoir," Winsor comments, "and shows the great taxonomic weight which Thompson gave to the larval forms."[8] Here we have our answer to the question about underlying explanations. For Thompson, embryology was so significant that he considered that larval (or embryonic) features eclipsed adult features in matters of classification. He was even prepared to deny that stalked and sessile barnacles (which resembled each other closely as adults) were near relations, simply because their larval features differed considerably. Embryology constituted the underlying explanation or epistemological stance that committed Thompson to the significance of the larval attributes.

Contemporary zoologists viewed Thompson's (correct) discovery that barnacle young exist in larval form in different ways, which reflected their own distinctive theoretical commitments. The new information about barnacles in no way clinched these creatures' classification.

Henri Milne-Edwards praised Thompson's work, grouping the Cirripedia with the Crustacea in 1852, when he revised his classification of the Crustacea.[9] But Milne-Edwards had previously shown himself willing to accept information from larval forms as indicative of taxonomic membership and, since 1829, had been formulating a theory which stressed the fundamental importance of embryology in determining class affinity. We can therefore conclude that for Milne-Edwards, as for Thompson, embryology acted as the underlying explanation or epistemological stance which highlighted the overriding significance of larval features for classification.

In 1835, Martin-Saint-Ange effaced the similarities earlier noted between cirripedes and mollusks by showing, on the basis of a detailed anatomical study, that earlier researchers had been wrong to consider barnacles unsegmented. His study illustrated that barnacles share more similarities with the Crustacea than with any other class and that, in common with the annelids, they are hermaphroditic. He thus placed the Cirripedia as a separate but transitional class between the Crustacea and the Annelida, paying no attention to Thompson's discovery in so doing.

Despite Martin-Saint-Ange's insistence that the cirripedes share more in common with the Crustacea than with any other class, he still placed them in their own category. Why? Perhaps his interest in the cirripedes' hermaphroditism explains why he placed them between the crustacea and the hermaphroditic annelids. But why did he consider hermaphroditism of such weight that it drew the cirripedes out of the Crustacea and into their own distinct class? By classifying the cirripedes separately, it seems that Martin-Saint-Ange could rectify what he considered an irregularity in Ampère's zoological system. Winsor tells us that Ampère "had arranged the animal kingdom into two parallel series, each of which ran from the simple to the complex, and which was symmetrical . . . each group in some way corresponded to the group at the same level on the other series."[10] Martin-Saint-Ange argued that grouping the cirripedes with the mollusks had not fitted this system, but according to his reclassification, the cirripedes now stood opposite the cephalopods (marine mollusks, characterized by well-developed head and eyes and a ring of sucker-bearing tentacles, such as octopuses, squid and cuttlefish), the correspondence consisting in possession of a soft body covered by a shell. This reclassification apparently wiped out the previous irregularity.

Winsor explains that systems like Ampère's were highly popular at this time, with many scientists favoring the overall aim of revealing a logical, regular pattern in taxonomies, even if they did not agree over specific details. It looks as if Martin-Saint-Ange, in tune with the time, wanted to show that systems such as Ampère's were well-founded and, by isolating the Cirripedia class and so apparently ironing out one of Ampère's irregularities, he came one step nearer to achieving this goal. Martin-Saint-Ange's epistemological commitment to Ampère-type systems explains why he considered the metaphysical property of hermaphroditism so salient that it compelled him to place the cirripedes in a unique, transitional class between the Crustacea and the Annelida. This epistemological stance ensured that Martin-Saint-Ange would count hermaphroditism—a property of the adult form—as far more important for classification than any property peculiar to the larval form.

Despite Richard Owen's belief that larval forms were very important for determining taxonomic position, still he did not concur with Thompson that cirripedes were crustacea. For him, locomotion was a vital characteristic, since he believed that increase in an animal's power of movement indicated progress. Although cirripedes possessed jointed appendages, these were used primarily for food gathering, he argued, rather than for the sake of locomotion. He, like Martin-Saint-Ange, therefore placed the Cirripedia as a separate class between the Crustacea and the Annelida. Owen later noted that although he considered the mobile larval cirripedes crustacean, once they lost their mobility as adults, they could no longer be considered members of the Crustacea class. We can therefore conclude that even though Owen considered embryology important, he considered locomotion more important still. His commitment to the connection between progress and locomotion underlay his rejection of the larval cirripedes' mobility in favour of the adult cirripedes' immobility as definitive for taxonomic decisions. Once more, a scientist's epistemological stance serves to highlight the significance of certain metaphysical properties over and above others.

Last, but by no means least, Charles Darwin was convinced that the cirripedes should be included in the class of Crustacea. He, naturally, was committed to his own theory of evolution, believing that those creatures descended from a common ancestor should be grouped together. Part of evolutionary theory declared that larval forms are the most likely to retain ancestral traits. We would therefore expect Darwin to consider (as he did) larval form and embryology in general as highly relevant for classification. Since the larval form of cirripedes was crustacean, Darwin insisted that the cirripedes belonged

	Dean	**Desmond**	**Winsor**
Classification not clear-cut	Orthodox vs. experimental taxonomy	Dinosauria vs. fossil lizards	Should larval properties influence taxonomic decisions?
Classification more than metaphysics	Morphological discontinuity vs. interbreeding/gene exchange	Mammalian similarities vs. reptilian similarities	Similarity between larval cirripedes and crustacea vs. similarity between adult cirripedes and annelids
Classification involves epistemology	Orthodox commitment to essentialism and ease/speed of classification vs. biosystematists' commitment to experimental methodology	Unquestioning commitment to established lizard paradigm vs. hatred of Lamarckism/Grant and favorable climate	Differing theoretical commitments influence perception of importance of embryology for taxonomy
Classification as synthesis of metaphysics and epistemology	Theoretical commitments highlight either morphological discontinuity or interbreeding/gene exchange	Theoretical/personal commitments highlight either mammalian or reptilian similarities	Differing theoretical commitments highlight either similarities between larval or similarities between adult forms
Explanation-based view applies to expert classification	Essentialism and ease/speed explains acceptance of morphological discontinuity Experimental methodology explains acceptance of interbreeding/gene exchange	Commitment to lizard paradigm explains acceptance of reptilian similarities Hatred of Lamarckism/Grant and favorable climate explains promotion of mammalian similarities	Differing theoretical commitments and perception of embryology explain acceptance of either larval or adult similarities

Table 1. Three case studies illustrating five theses.

7: Five Interrelated Theses

in the Crustacea class, regardless of the dissimilarity between crustacea and adult cirripedes. For Darwin, similarities involving the larval form were conclusive, isolated by his commitment to evolution as an underlying explanation.

And so we see that Winsor's case study again illustrates the complexity of the natural world together with the different patterns of similarity that cut across that world. The cirripedes really do share properties in common with both the crustacea (similar jaws and jointed limbs in the larval form) and the annelids (segmented bodies and hermaphroditism). Metaphysics alone does not reveal a uniquely correct taxonomic position for the cirripedes. Rather, whether individual scientists interpreted the cirripedes' larval features as relevant for classification depended upon their underlying theoretical (epistemological) allegiance. Differing allegiances produced differing classifications, all of which relied on real, shared similarities. Such metaphysical and epistemological features combine to provide the wider arena in which categorization must take place, an arena in which—due to the inherent richness of the natural environment—controversy and difference inevitably arise.

To recap, I used three case studies to illustrate five interrelated theses concerning scientific classification, shown in Table 1.

These three case studies illustrated both the benefits and the mechanics of my pluralist account of scientific classification. We have seen that the pluralist position quite naturally incorporates the kind of disputes that arise over classification, since it starts out with the notion of a complex, multi-patterned environment. This account allows that epistemology plays an important part in classification and allows that competing classifications based on real properties may all be correct. In each of these respects, my position challenges Objectivism. It also succeeds where Objectivism fails, since it alone accounts for the differences that have existed—and continue to exist—within scientific taxonomy.

8
Philosophical Contexts I: Critiques of Natural Kinds

My pluralist account of scientific classification does not, of course, stand isolated, but falls in with a philosophical tradition. In particular, I share assumptions and arguments in common with those thinkers who have criticised Putnam and Kripke's rigid realist approach to natural kinds and essentialism. I want now to take a look at the work of two such philosophers—Keith Donnellan and John Canfield. We will see that they, like me, argue that division of the natural world into kinds involves scientific decision-making and is not determined by nature alone. They also argue that two competing definitions of a particular natural kind may attain truth. By examining their positions, we will provide a context for my account of scientific classification and will come to understand how my work relates to that of other contemporary philosophers.

Keith Donnellan's Twin Earth

Donnellan launches an attack on Putnam's notion of important physical properties, which are said to be discoverable by science and so to provide the true definition and demarcation of natural kinds. The heart of Putnam's objectivist position is conveyed by the following two quotations:

> To be water, for example, is to bear the relation same$_L$ [same liquid] to certain things.

> x bears the relation same$_L$ to y just in case (1) x and y are both liquids, and (2) x and y agree in important physical properties.[1]

And the point he aims to convey—using water as a specific example—is that nature divides inherently into certain (natural) kinds and these kinds are underlain by possession of certain fundamental, defining, usually microstructural properties. Once these properties have been discovered (by science), there is only one way in which we can *correctly* divide up the natural world—the way which reflects these (natural) kinds and their defining properties.

But Donnellan is sceptical about Putnam's account, illustrating its failings via the following thought experiment. He asks us to imagine two cultures, one on Earth (and so identical with our own culture) and one on Twin Earth which, up to a certain point in their respective histories, are absolutely alike in every possible way. In particular, their languages are identical and they both developed identical sophisticated sciences, so embracing scientific viewpoints of the world. They both developed atomic theory. Both, therefore, have the concept of an atom as having a nucleus which comprises positively charged particles (protons) and neutrally charged particles (neutrons). They both have the concept of elements as non-compound substances whose atoms have a particular number of protons in the nucleus and they both call this number atomic number. They both have the concept of isotopes as being individuated by the combined number of protons and neutrons in the nucleus of the atom and they both call this the isotope number. They are both aware that one and the same element usually has several different isotopes and they both give separate names to each isotope of the element.[2]

On Twin Earth, it is generally the case that one of the isotopes of a particular element comprises the bulk of the naturally occurring element, while the remaining isotopes are pretty rare.[3] And (on both Earth and Twin Earth) different isotopes of the same element display important differences in behavior. Some isotopes are radioactive, while others of the same element are not; some isotopes are unstable and break down into isotopes of other elements, while others of the same element do not. It will, of course, also be significant which *element* we are dealing with, particularly in relation to chemical reactions.

Donnellan's point is this: it is not at all clear on the Earth/Twin Earth scenario whether atomic number or isotope number is the more fundamental hidden property, so far as those substances that we call the 'chemical elements' are concerned. The problem, of course, centers on Putnam's insistence that scientists isolate *the* important property or properties underlying a specific natural kind. Sometimes—contrary to Putnam's beliefs—scientists can discover *more than one* important hidden property. The issue then becomes one of deciding which property will count for classification or natural kind division. And this is a crucial issue, since different defining properties will result in different natural kinds.

Both atomic number and isotope number are important physical properties in the sense that Putnam's account appears to require—they are both associated with the microstructure of substances. Each property also accounts for certain uniform behaviors—radioactivity, breakdown and chemical reaction—which take place within or between particular substances. It is therefore arguable that both atomic and isotope number represent equally good candidates for *the* important property, the one which is relevant for dividing the natural world and so determining kind membership. Since the natural world appears ambiguous here, we may argue—in my terminology—that classification must involve more than mirroring metaphysics or the way the world is.

Donnellan applies the details of his story to specific examples. It is not implausible, he claims, that on Twin Earth 'gold' is used to label the isotope which makes up the bulk of the element with atomic number 79, rather than simply the element with atomic number 79, as on Earth. Twin Earthians would then fail to consider the rarer isotopes true gold, even though they bore strong similarity to gold in certain respects, while Earthians would make no such distinction.

On Putnam's account of natural kinds, it therefore becomes possible that with two cultures sharing exactly the same linguistic and scientific background, a particular kind term could have a different extension in each culture. But this is a possibility Putnam would wish to avoid. For him, natural kinds are just that—natural—they represent the inherent structure of nature and so by definition have only one possible extension. So Donnellan argues that we cannot agree with Putnam that natural kind terms have the same extension before and after scientific discovery of their hidden properties. Which way scientists *choose to interpret and apply* these discoveries will affect the extension of those terms. Suppose, he says, that John Locke possessed a ring made out of one of the rarer isotopes of what we call gold. Locke would have claimed, on the basis of its surface properties, that his ring was gold. Post-scientific discovery, his claim would remain correct for Earthians, but become incorrect for Twin Earthians. On Twin Earth, the extension of 'gold' has actually changed *as a result of* scientific discovery due to isotope rather than atomic number being taken as definitive of the natural kind. Donnellan therefore concludes that going part of the way with Putnam forces us to "admit nature does not fully determine the extension of vernacular natural kind terms and science is not wholly responsible for discovering their true extensions."[4]

Since—as Donnellan shows—there can be more than one suitable candidate for defining a natural kind, scientists must make a choice or decision regarding which property is to be taken as definitive of the kind. Scientists do not simply reflect the hidden structure of the natural world, but actually play an active part in determining the nature of natural kinds.

In the Earth/Twin Earth example, Earthian scientists have, at some point in the past, made the decision that what is to count as the important physical property for delimiting the chemical elements is atomic number. Twin Earthian scientists, on the other hand, have decided that isotope number should count for dividing chemical substances into kinds. Presumably any number of factors may prompt the choice of one property over another. In reality, atomic theory was accepted prior to the discovery of isotopes. Perhaps this influenced the way in which we mapped vernacular kind terms onto science. But, as Donellan argues, timing is a mere contingency—it tells us nothing decisive about the way the world is. We might equally imagine that chemical substances were originally divided according to atomic number on Twin Earth, but that following the discovery of isotopes, Twin Earthian scientists considered division according to isotope number constituted a more precise system which better suited their purposes. Or we can imagine that isotope number was discovered on Twin Earth prior to atomic number. Or perhaps that both were discovered at roughly the same time, but that for some reason, isotope number was adopted in preference to atomic number. The motivations need not concern us. What we *do* need to take on board, however, is Donnellan's confirmation that scientists may make use of different (real) properties in defining natural kinds. This means that human beings play an active role in the classificatory process and so classification involves a synthesis of epistemology (human theorizing, decision and choice) and metaphysics.

The Earth/Twin Earth story obviously results in differing truth values for certain sentences, dependent on whether one is Earthian or Twin Earthian. Donnellan lists the following examples. "Some gold has isotope number x and some has isotope number y," will be true for Earthians, but false for Twin Earthians. "Gold has atomic number 79," will be true for Twin Earthians, so long as the word 'has' does not involve any notion of identity, but will be straightforwardly true (by identity) for Earthians. And "Gold is identical with the element with atomic number 79," will be true for Earthians and false for Twin Earthians.

Does this difference in truth value mean that we must judge one natural kind definition right and the other wrong? Of course not. How could we possibly judge one right and the other wrong? Nature does not provide a single, conclusive answer, as Putnam hoped. Indeed, it is nature's very ambiguity which allows for the setting up of the thought experiment in the first place—and it is this ambiguity which renders Earthian and Twin Earthian definitions equally correct. Since both highlight similarities or regularities which really are 'out there' in the natural world (similarity of chemical reaction with atomic number and similarity in radioactivity and stability/breakdown with isotope number), we must judge them equally objective, legitimate and correct.

John Canfield's Criterial Theory of Essence

Canfield presents a theory to rival Kripke and Putnam's realism regarding natural kinds. This theory rests on the notion that essence is not *de re*, but *de dicto* and is couched in terms of a Wittgensteinian criterial account whereby essences are determined not by reality—as Putnam and Kripke insist—but by criterial rules of language.

He begins by distinguishing between intentional and extensional discoveries. Intentional discoveries are 'discoveries that. . .'. We might, for example, discover intentionally that the essence of gold is atomic number 79. All other discoveries are extensional. So, rather than discovering that the essence of gold is atomic number 79, we might simply uncover extensionally a number of important properties which gold possesses, including its atomic number.[5] Canfield then asks: "But does the scientist merely discover that gold has the atomic number, thereby discovering essence; or does he, alternatively, discover *that* it is the essence of gold to have this number."[6]

Kripke argues for the second intentional alternative, Canfield suggests—by discovering that gold has atomic number 79, scientists in fact discovered that it was the essence of gold to possess atomic number 79.

But Canfield cannot accept Kripke's interpretation. How can we discover the essence of something *per se*? How would we know that a particular property or properties were the essential ones? We have discovered all sorts of scientific properties of gold—its lattice constant, its melting and boiling points, its coefficient of linear expansion, its tensile strength, its Brinall hardness, that it is a face-centered cubic metal—yet it is unreasonable to claim that all are essential. Once again, we see that the world is ambiguous with regard to natural kinds and their essences. Although atomic number 79 has been accepted as the essence of gold, various alternative properties might—equally legitimately—have fulfilled this role. The properties that Canfield lists are all scientifically relevant properties which capture important aspects of the substance we call 'gold.' They all satisfy, for example, Putnam's criterion of being important physical properties.

Canfield solves this puzzle by concluding that some kind of *theory of essence must be presupposed* before we can move from the claim that gold has atomic number 79 to the claim that the essence of gold involves being an element with atomic number 79. It is the theory that then enables us to isolate essential from accidental properties. Canfield's theory of essence bears a striking resemblance to my own argument, detailed in the previous chapter, that the explanation-based model of categorization (which accounts for the choice of specific properties via an underlying theory or explanation) characterizes the way in which scientists divide natural entities into kinds. We both seem to concur, then, that scientific classification involves a synthesis of metaphysics and epistemology.

Kripke employs a strategy for defending his account of natural kind terms, which involves constructing abnormal states of affairs and then asking, for instance, whether these particular objects are tigers or whether this particular substance is gold.[7] The answers we give to these sorts of question—which arise from our intuitions—go against criterial accounts, Kripke argues, hence criterial features cannot determine kind membership.

But Canfield remains unconvinced. He believes we can construct Kripke-type examples where our intuitions give results that *conform* to the criterial view and disconfirm Kripke's own theory. He invites us to imagine that animals looking very like what we call tigers are dubbed 'tigers' by explorers—the explorers thus introduce the term into their own language. People who subsequently come into contact with these animals also consider that they possess tiger-like characteristics. But in reality, an illusion surrounds the animals and, under veridical conditions, they resemble what we call chickens. What happens if the illusion clears? The answer is not clear-cut, Canfield claims. We might say that there are, after all, no tigers in this case, but we might equally say that there are tigers, but they turned out to be different from how we originally supposed. (Kripke would, of course, argue that only the second statement attains truth, since he considers surface, criterial features irrelevant for determining kind membership.) Each response is perfectly understandable, Canfield continues, and furthermore, each is true.

How is it possible that both responses attain truth? Contra Kripke, Canfield argues that our intuitions about such matters will arise from presuppositions involving the choice of a particular criterion for extension of the normal use of kind terms to abnormal situations—but there will also be alternative criteria which we could equally legitimately presuppose. If we claim there are no tigers, we are adhering to the criterion that something is a tiger only if it possesses at least some of a collection of specified features. If we claim that there are tigers, we are adhering to the alternative criterion that something is a tiger if it is the entity pointed to in the original reference-fixing act or if it lies behind our original misperceptions. In other words, we must make a choice. Should we define natural kinds on the basis of their surface features or should we define them on the basis of some kind of essential, inner or microstructural properties? Either choice is legitimate, according to Canfield. But, once the choice is made, it becomes concrete—a presupposition or intuition which guides our future thinking about natural kinds.

Kripke's theory then fails because it presupposes what it aims to prove. "Everything depends on what we mean by, or count as, 'the same kind of stuff,'"[8] and by insisting (for example) that two samples only count as gold if they both have atomic number 79, Kripke already presupposes that atomic number 79 is essential for determining that the two samples are of the same (kind of) stuff—gold. He jumps the crucial step of criterion choice—the step that I have dubbed adopting an epistemological stance—instead presupposing a particular criterion and taking that as unquestionably correct. We might add that Kripke (and Putnam) concentrate on *currently accepted* natural kind definitions and work backwards from there.[9] Such an approach will, of course, yield the impression that nature unambiguously fixes natural kinds and their essences, since we tend to assume, as I have mentioned in previous chapters, that either the accepted definition represents the one and only correct definition, or that such a definition exists but still awaits discovery. As Canfield emphasizes, "The natural use of "natural kind" does not carve up the conceptual territory in the way that Kripke wants it carved up, and if we want to use "natural kind" to carve it up that way we shall have to put a stipulation on its use."[10] We will have to make a decision about which properties are to count as essential for natural kind membership. And so Canfield's account confirms my claim that epistemology in conjunction with metaphysics defines and delimits the kinds we use.

How, then, does the criterial view of kind terms account for scientific discovery of essence? Canfield explains with reference to the discovery and definition of viruses. Historian of science Sally Smith Hughes tells us that, following the discovery of these entities, the problem for scientists "was to discover the intrinsic properties of viruses rather than to characterise them in terms of technique determined ones."[11] Reading 'essential properties' for Hughes' 'intrinsic properties,' it looks as if we have a Kripke-type situation in which viruses were initially isolated on the basis of extrinsic or surface features with scientists later uncovering their underlying essence. Hughes continues:

> By the 1950s, structural studies of viruses with the electron microscope and information about their nucleic acid content provided a meaningful basis for distinguishing viruses from all other types of infectious agents. From this time on, 'virus' was used with a meaning roughly comparable to that given it today. Modern definitions abound but most characterize the virus as an infectious, but not necessarily pathogenic, entity which is usually submicroscopic, which contains a core of either DNA or RNA covered by a protein or lipoprotein capsid and which reproduces exclusively within living cells.[12]

This statement fits the criterial account just as well as the rigid realist one, Canfield argues. How? Scientists, on examining viruses, discover a number of properties that these entities were not previously known to possess. Some of these properties will be used in constructing a definition or definitions of a virus. But it is *only when a definition is provided and accepted* that the essence of a virus emerges, not before. Hence scientists do not intentionally discover that *x* is the essence of viruses, but stipulate that a certain (extensionally discovered) property or properties are *to be counted as* the essence of viruses. In doing so, they make an active contribution to the definition of viruses. They must make a choice between potential essential properties, which reinforces my position that the adoption of an epistemological stance is necessary in order to get a particular classification up and running. Canfield here highlights Hughes' statement that there are a number of different extant definitions of viruses. And he argues that when the term 'virus' is used with a particular definition in mind, then, by dint of that very definition, the essence of what it is to be a virus *in that context* is invoked. However, if the term is used with a different definition in mind, then the essence associated with that alternative definition is invoked. Which definition we favor determines which (metaphysical) properties are to count as relevant for natural kind membership.

This example allows for the wider possibility that scientists have a hand in stipulating not only the essence, but also the *boundaries* of natural kinds. Scientists may, for example, discover that all known viruses have an inner core of either RNA or DNA, but still decide not to include this fact in the definition of viruses. If, as an alternative, they decide to include being of a certain size and possessing a specified inner molecular structure within the essence of viruses, they will, by that very decision, ensure that in future they never discover a *virus* without those properties—although they may discover an entity of similar size and molecular structure which displays similar behavior. Only if they subsequently decide to change the definition, might this entity be considered a virus. This is not, of course, to say that scientists can adopt any definition they like. They will always be constrained by the real, objective properties that careful scientific discovery reveals—such as that certain microscopic entities possess inner cores of DNA and RNA. But, given the multiplicity of such (real) properties, scientists must make a choice regarding which one(s) they will use in defining the class of viruses. And which one(s) they choose will (partially) determine the boundaries of the class. "This illustrates the element of decision involved when a term is defined by means of properties science has discovered," Canfield concludes.[13] In other words, scientific classification comprises a synthesis of epistemology (decision-making) and metaphysics (real, scientific properties).

We have seen that several points of contact exist between my pluralist account of scientific classification and critiques of the Putnam/Kripke approach to natural kinds forthcoming from philosophers like Keith Donnellan and John Canfield. In particular, Donnellan, Canfield and I all share the belief that nature alone cannot completely determine either natural kinds or their essences. We therefore agree that scientists must, by choosing certain properties over others, play an active role in determining both the essential nature and the boundaries of these classes. We would also all, I think, label ourselves realists, since we believe scientific classification and demarcation of essence must always be constrained by real, mind-independent properties exhibited by real, mind-independent entities. While we acknowledge that active, human decision-making is a necessary part of scientific classification, we aim to abandon not realism, but only that brand of unrealistically rigid realism touted by objectivists like the early Putnam and Kripke.

It is possible, then, to situate my pluralism in the philosophical landscape alongside critiques of the objectivist understanding of natural kinds. But my views can also be situated with those of certain contemporary philosophers of biology. It is to these thinkers that we turn in the next chapter.

9
Philosophical Contexts II: Contemporary Philosophy of Biology

In chapter 7, I mentioned that the debate between orthodox and experimental taxonomists continues today in modified form as biologists and philosophers of biology seek to impose competing classificatory schemes upon the natural world. In fact, three broad schools currently vie for supremacy—phenetics (also known as numerical taxonomy), evolutionary systematics, and phylogeny (or cladism)—and each generates a variety of different species concepts. Proponents of these competing schools seek to establish their own system as supreme and all-purpose, denigrating the others as inadequate in various respects. The situation remains unresolved and it seems unlikely, on metaphysical grounds, that a single school will ever gain supremacy over its competitors. As a result, some philosophers have recently moved beyond the current impasse, arguing that no single, unique form of classification can exist, but that different schools of taxonomy reflecting different perspectives on, and interests in, the natural world can and should coexist.

The aims of this chapter are three-fold. First, I discuss the three current schools of taxonomy, emphasizing both their strengths and weaknesses.[1] We see that the resulting debate over species concepts is a live and very real one, not something we can simply relegate to the history of science. Second, we see that the disparity between species concepts should not be understood as problematic, but actually reveals the inherent complexity of the environment and allows for a plurality of scientific perspectives on that environment. And third, I discuss Philip Kitcher's, John Dupré's and Ian Hacking's theorizing about natural kinds, indicating the similarities and differences between their positions and my own pluralist account of scientific classification.

9: Philosophical Contexts II

Phenetics

Phenetics (or numerical taxonomy) is a school of morphologically-based taxonomy which grew up in the 1960s and 1970s and is most commonly associated with Robert Sokal and Peter Sneath.

The basic notion behind phenetics is that biological organisms should be classified according to *overall similarity* (incorporating similarity of function, form and biological role), calculated using numerous characters. Each character used in calculating overall similarity is given equal weight—no character is preferred over any other. This work is mathematical—each character is first recorded in numerical form and overall similarity is then calculated by algorithmically manipulating the characters with the aid of a computer to produce a diagrammatic chart of the relationship between the entities to be classified, known as a phenogram. This diagram of phenetic distances is then converted into a biological classification. The name 'numerical taxonomy' reflects the pheneticists' mathematical bias.

Phenetics aims to produce uniform and objective classifications, and adherents to this school believe that such classifications can be produced by explicitly formulated statistical methods. Pheneticists are of the opinion that overall similarities—there for one and all to see—must stand prior to *a priori* speculation concerning biological process. David Hull reports that in phenetics, "Classifications should unequivocally represent the resemblances exhibited by organisms without regard to their descent. Descent could then be indicated in an accompanying diagram."[2]

We can construe phenetics as a revival (in more respectable and modern form) of the orthodox system of taxonomy based on morphological discontinuity, which we discussed in chapter 7. Yet, despite the enthusiasm and commitment of its original proponents, phenetics has largely fallen into disfavor. Although it is still respected in areas such as microbiology and botany, biologists in general acknowledge that some principle—independent of similarity—is necessary for judging which morphological features are and which are not relevant for classification. "This seems necessary on philosophical grounds," explains John Dupré, "to avoid the difficulties with making sense of absolute similarity, and on biological grounds to assure that properties selected have suitable evolutionary or other theoretical significance."[3] Throughout the next few sections, we will see how important the theme of evolution has become for taxonomy.

Evolutionary Systematics and the Biological Species Concept

The biological species concept—typically associated with the work of Ernst Mayr—defines species as: "Groups of actually or potentially interbreeding populations which are reproductively isolated from other such groups."[4]

The basic notion here is that members of a species together constitute a reproductive community and so respond to one another as potential mates. The species also constitutes an ecological unit which interacts as a whole with other species occupying the same ecological area. And finally, the species forms a genetic unit which constitutes a sizeable interacting gene pool, in contrast to an individual member of the species which "is merely a temporary vessel holding a small portion of the contents of the gene pool for a short period of time."[5] Again, we see the similarity with biosystematics—discussed in chapter 7—which individuated species on the basis of interbreeding and gene exchange.

Mayr emphasizes that a species is a *protected* gene pool which has so-called "isolating mechanisms" protecting it from the potentially harmful flow of genes from other pools. Genes from the same pool combine harmoniously because they have become adapted to one another as a result of natural selection. But the mixing of genes from different pools will result in disharmonious combinations, hence mechanisms preventing such mixing are favored by natural selection.

The biological species concept, then, (unlike phenetics) rests upon the presupposition of an inextricable link between taxonomy and evolution. Mayr in fact states that, "An understanding of the nature of species . . . is an indispensable prerequisite for the understanding of the evolutionary process."[6]

Despite its apparent simplicity, the biological species concept does entail a number of limitations and problems. Although the majority of species comprise aggregates of numerous local populations which exchange genes, the further away populations are from one another, the more likely it is that their respective characteristics will differ. Mayr lists the possible differences: widespread species may have terminal populations which do not interbreed even though contiguous populations in the chain do interbreed; populations may become reproductively isolated while remaining morphologically very similar;

populations may diverge morphologically yet continue to interbreed; populations may quite happily interbreed yet, when their *habitats* change, they may become reproductively isolated; isolating mechanisms are built up slowly over time and so can be imperfect and incomplete; isolating mechanisms are established at different rates in different populations, hence it is possible that in some areas two populations are reproductively isolated, while in others, they interbreed.[7]

How, then, can we determine in these problem cases what is and what is not a species according to the definition of the biological species concept? No concrete answer presents itself and Mayr admits that, "Determination of species status of a given population is difficult or arbitrary in many of these cases."[8]

A second problem concerns asexual organisms. By definition, the biological species concept cannot apply to such individuals and so we require some other set of criteria to determine species membership in these cases. Typically, adherents to the biological species concept utilize clear discontinuities—produced by mutations in the asexual lines—to delimit species among asexual organisms.

Thirdly, Sokal and Theodore Crovello claim that the biological species concept suffers from "phenetic bottlenecks."[9] By this they mean that, due to time limitations, all users of the concept will initially have to group individuals and populations on the basis of phenetic similarities rather than patterns of interbreeding. While the definition of the biological species concept makes no mention of phenetics, its determination will, of necessity, always involve phenetic considerations. Even where crossing tests have been performed, Sokal and Crovello claim that the basic species definition is still phenetic since any subsequent statements are based on *phenetic* inferences from the small number of crosses actually performed. They therefore conclude that the biological species concept does not form a necessary part of evolutionary taxonomy, since most of the evidence for evolutionary taxonomy rests not on interbreeding, but on phenetics.

Sokal and Crovello's argument may sound convincing at first blush, but I suggest that it does not stand up to analysis. Their claim—that the biological species concept is not a relevant part of evolutionary taxonomy because most of the evidence for evolutionary taxonomy is based on phenetics and not interbreeding—misses an important point. The only reason evolution or interbreeding is salient at all is that we are focusing on the *biological* species concept—and it is the *biological* species concept which aims to reflect elements of the process of evolution by concentrating on interbreeding of individuals and populations. If we abandon the biological species concept, we abandon along with it our interest in evolutionary taxonomy, unless we replace it with another species concept which invokes an *evolutionary*-based taxonomy—which phenetics does not. Sokal and Crovello's mistake is to confound the *principle* behind the biological species concept (reflection of the link between evolution, interbreeding and species) with the *process* by which that principle is achieved (which may involve use of phenetic similarities for speed and ease). Their insistence that we are left with "what is essentially a phenetic criterion of homogeneous groups that show definite aspects of geographic connectedness" similarly misses the point.[10] What (essentially) motivates a phenetic criterion is the desire to produce a classification based on all observable properties possessed by an organism, where *all* these properties are given *equal* weight, whereas what (essentially) motivates the biological species concept is a desire to reflect patterns of interbreeding between organisms. What we are left with, then, is anything but an *essentially* phenetic criterion.

Phylogeny

Several slightly differing methods of classification fall under the heading of phylogenetic species concepts.[11] There is also some ambiguity over terminology—systematists sometimes talk of phylogeny, while at other times they talk of cladism—but the two terms refer (at least roughly) to the same (group of) theories.

The central tenet underlying (earlier accounts of) phylogenetic classification is that classifications should reflect genealogy or evolutionary branching patterns. In order for a group of organisms to form a species, they must share some kind of common ancestry. Phylogeny posits a direct link between taxonomy and the history of evolution. In the words of Kevin de Queiroz, "Phylogenetic definitions are thus firmly rooted in the concept of evolution, that is, of common descent. . . . What is both necessary and sufficient is being descended from a particular ancestor."[12]

Willi Hennig is perhaps the best known proponent of cladism. He believed that traditional hierarchical classifications were not complex enough to reflect all the details of phylogenetic development. He therefore concentrated on one particular element in phylogeny—the sister-group relationship. Two taxa (A and B) represent a sister group if they are more closely related to one another than they are to any other taxon (C), the proximity of relationship being based on characters which A and B share with one another but do not share with C. The sister-group relationship is collateral (not ancestor-descendant), hence A and B must share a more recent common ancestor with one another that either one of them does with C. None of the taxa within the statement of a sister-group relationship are said to be ancestral to any other. Despite the fact that Hennig recognized two types of phylogenetic relationship (the sister-group and ancestor-descendant relationships), he still insisted that a truly phylogenetic classification concerns sister-group relationships alone.

According to Hennig's account of cladism, the sister-group relationship can be represented in a phylogenetic diagram (or cladogram), in which speciation events are represented as the splitting of a single line (the stem species) into two (the daughter species which form a sister-group). Hennig argued that when the ancestral species splits, we must consider it extinct, and only when this splitting occurs should we recognize new species.[13] Thus an ancestral species cannot exist alongside its descendants and two concurrently existing species can only be connected by the sister-group relation, never by the ancestor-descendant relation.

Hennig also considered that taxa should be monophyletic. Ordinarily, this means that a taxon should include only the ancestral species and its descendants, but Hennig extended this, arguing that a taxon should include the ancestral species together with *all* its descendants. This limitation has some surprising consequences, as we can see in the following examples.[14] Crocodiles and lizards retain many of the features of their common ancestor and are highly similar, but crocodiles and birds have a more recent common ancestor which is not an ancestor of the lizards. Hennig's cladistic classification would therefore group crocodiles and birds together since they form a genealogical unit apart from lizards. Secondly, placental and marsupial wolves independently evolved a number of similar characteristics. However, placental wolves form a genealogical unit with moles. This means that the Hennigian cladist would group moles and placental wolves together, but exclude marsupial wolves.

Despite Hennig's somewhat radical claims, John Dupré notes that, "Probably a majority of those sympathetic to phylogenetic taxonomy are in fact committed only to the much weaker demand that classification not be inconsistent with the genealogical tree."[15] This more conservative attitude will generally require that all the members of a taxon be monophyletic but will not necessarily require that the taxon include the ancestral species together with *all* its descendants. A more conservative position would therefore classify crocodiles and birds separately (as has traditionally been the case), motivated by the belief that taxonomy should, to a certain extent, reflect similarity and difference independent of phylogeny.

A more conservative approach to phylogeny also allows for the possibility of speciation *without* splitting (anagenetic speciation), where enough change within an undivided lineage seems to warrant demarcation of more than one species. One example of such an approach is Gareth Nelson and Norman Platnick's pattern cladism. Pattern cladists define species as "simply the smallest detected samples of self-perpetuating organisms that have unique sets of characters."[16] They abandon Hennig's claim that we can only distinguish separate species on the basis of speciation events, instead urging that cladistics should focus on "detectable changes."[17] Nelson and Platnick try to develop a more general account of cladism than that provided by Hennig, arguing that terms such as 'cladism' and 'cladogram' "unfortunately were intended to have an explicitly evolutionary significance pertaining to the actual branching or speciation events of phylogeny,"[18] and stating that under their alternative account, "Cladograms depict structural elements of knowledge."[19,20]

As was the case for both phenetics and the biological species concept, systematists have perceived advantages and disadvantages with cladistics.

Joel Cracraft suggests that the phylogenetic species concept improves on the biological species concept in a number of ways.[21] Phylogenetic taxonomic units are, by definition, equivalent to known evolutionary units, which leads to greater clarification of evolutionary issues. The phylogenetic concept clarifies the distinction between recognizing species taxa and analyzing geographic variation. According to phylogeny, subspecies cannot have ontological status as evolutionary units. Cladists do not use data about reproductive isolation to demarcate species and so even where two sister-taxa hybridize, each is still recognized as a separate species. Phylogeny focuses more attention on the geographical history of a species that does the biological species concept. Cladism places strong emphasis on the search

9: Philosophical Contexts II

for diagnostic characters to be used in delimiting taxa, rather than concentrating on the description of variation within taxa. Since the phylogenetic species concept aims to identify all evolutionary taxonomic units, it produces a much more accurate assessment of inter- and intra-cladal diversity patterns than does the biological species concept.

Cracraft also indicates a possible disadvantage. The phylogenetic species concept recognizes many more species level taxa than its biological counterpart, thus eschewing Mayr's aim of introducing greater clarity and simplicity into biological classification. Pheneticists point out a different disadvantage. Although Sokal and Sneath admit that phylogeny is "an all-explanatory principle," they argue that it should not be used in classification "since we mostly do not know (and in many cases cannot know) its true course."[22]

✳

We have seen that, even today, systematists are far from agreeing on a universal, all-purpose school of taxonomy. Contemporary biological theory fails to yield anything like a system with which we might produce a unique classification of the biological world and so satisfy adherents to Objectivism. Instead, we have three different schools of taxonomy—phenetics, Mayr's biological species concept, and various formulations of cladism—each of which is currently in use (albeit in limited areas only, in the case of phenetics) and each of which entails both advantages and disadvantages. This widespread disparity in biological taxonomy has prompted two philosophers—Philip Kitcher and John Dupré—to promote radical forms of pluralism which acknowledge the variety of existing species concepts and deny that any one merits precedence over its competitors.[23] How, then, does their species pluralism relate to my pluralist account of scientific classification?

Kitcher's pluralistic realism "rests on the idea that our objective interests may be diverse, that we may be objectively correct in pursuing biological enquiries which demand different forms of explanation, so that the patterning of nature generated in different areas of biology may cross-classify the constituents of nature."[24] I could not agree more. In chapter 7 I similarly argued that, although scientists produce competing classifications, still these competing classifications may all be correct. What makes a classification correct, I suggested, is that it reflects or corresponds to an important aspect of the real world.

Yet those scientists who champion a particular species concept tend to believe that their concept covers the needs of all biological taxonomists. Such a belief, Kitcher argues, is mistaken. So-called problems associated with the biological species concept, for example—that we cannot evaluate whether two long-extinct forms were reproductively isolated and that we cannot apply the concept to asexual organisms—must be taken seriously and not simply dismissed as infrequent anomalies. Why? Although the biological species concept concentrates on an important, theoretically significant pattern in nature—reproductive isolation—still this is not the only important, theoretically significant pattern available. Apparent problem cases "point to distinctions among organisms which can be used to generate alternative legitimate conceptions of species."[25]

How is such pluralism possible? Kitcher distinguishes between *structural explanation*—which explains the properties exhibited by organisms in terms of underlying mechanisms and structures—and *historical explanation*—which isolates the evolutionary forces which have produced the morphology, behavior and ecology of extant and extinct organisms. Neither type of explanation (nor the answers it produces) is more important or ultimate than the other, Kitcher insists. Each simply seeks to explain a *different* set of phenomena. The two kinds of biological explanation therefore produce differing schemes for the classification of organisms, one based on structural traits and the other based on historical (evolutionary) traits. Kitcher's analysis is right here, I believe. As we saw in chapter 7 when discussing the competition between orthodox and experimental taxonomists, different sets of traits generate different classification schemes. We cannot judge one scheme right and the other wrong because each scheme serves a different purpose (speed and ease of classification versus emergence and success of experimental methods) by concentrating on alternative sets of properties (morphological discontinuity versus patterns of interbreeding and gene exchange) occurring in the real world.

But Kitcher's pluralism runs deeper than this. Within each of the two schemes (structural and historical), there exist finer distinctions still. It may be possible to explain diversity among organisms by referring to structural traits at *different* levels. Explanations involving genetics may be appropriate to some organisms, for example, while explanations involving chromosomes may be more appropriate to other organisms. Kitcher illustrates this with the phylogenetic species concept. In order to get this concept off the ground, he explains that biologists need to employ a principle of phylogenetic division with which they can judge what changes are important enough to give birth to a new evolutionary unit. Three main candidates exist—the production of reproductively isolated branches, the attainment of ecological distinctness, and the development of a new morphology—and which principle of division a biologist adopts will (in part) depend on his/her area of interest and the nature of the organisms with which s/he is dealing. A paleontologist who is reconstructing the phylogenies of classes of organisms may therefore use the reproductive isolation of descendant branches to divide well-understood vertebrates into species, but may use ecological or morphological discontinuities to divide asexual plants or marine invertebrates into species. Here Kitcher tacitly implies what I have openly asserted—that biologists have a choice open to them. Should they, in Kitcher's example, concentrate on reproductive isolation, ecological distinctness or new morphology? The world is neutral—all three options represent biologically significant, naturally occurring phenomena; all three represent legitimate principles according to which the biologist might group organisms. "We see that organisms are so diverse that their diversity demands a diversity of diversities," Kitcher concludes.[26] I agree. The environment's inherent richness produces a situation in which biologists must choose between several objective, correct, yet conflicting principles of division.

Dupré's position closely resembles that of Kitcher. His promiscuous realism entails that, "There is no God-given, unique way to classify the innumerable and diverse products of the evolutionary process."[27] Darwinian theory of evolution forces us to realize that nature does not possess a unique structure, Dupré explains, and so our reasons for accepting a particular taxonomic scheme must be that it serves a particular purpose better than other possible schemes, not that it is *the* uniquely correct scheme of classification. "It is that the complexity and variety of the biological world is such that only a pluralistic approach is likely to prove adequate for its investigation," he continues, insisting that he is ontologically (and not methodologically) motivated to advocate pluralism.[28] Promiscuous realism, like both Kitcher's pluralistic realism and my pluralist account of scientific classification, rejects the objectivist notion that a single system of classification can satisfactorily capture the complexity of the natural world. This complexity is (to use my terminology) a matter of metaphysics and not of human construction, but the fact that such complexity exists means that biologists can classify organisms in different ways, thus reflecting different real aspects of the biological environment.

Does this kind of approach lead to relativism? No. Nothing about promiscuous realism forces us to abandon realism, since realism regarding biological kinds remains quite separate from the belief that only one factor exists with which we can account for diversity among biological organisms. Although recently favored species concepts (phylogenetic and biological species concepts) focus on the fundamental relevance of evolution for classifying organisms, Dupré indicates that there is more to biology than evolutionary theory. Ecology also provides an important theoretical context in which species play a part, he argues, since the application of ecological theory to real-life ecological situations depends on classification of the relevant organisms. As the occupants of a specific ecological niche do not always coincide with the members of a specific genealogical line, we can expect ecologists to prefer classificatory systems which do not concentrate on genealogy. It thus becomes possible for an organism to "belong to both one kind defined by a genealogical taxonomy and another defined by an ecologically driven taxonomy."[29] Dupré and I hold the common belief that both of these classifications are correct, since each one corresponds to a different, but important, aspect of reality. Evolution and ecology simply represent different epistemological starting points. These starting points serve to highlight alternative metaphysical properties and so produce competing but equally legitimate species concepts.

Ian Hacking, although not motivated by contemporary debates about biological species concepts, does hold views about natural kinds which relate to both my own position and that of Kitcher and Dupré. He considers that his view inclines towards nominalism, but is realist in recognizing that kinds arise in nature. It is a fact of *nature* that there exist differences among things in virtue of which they fall into different kinds, but recognition of these differences and kinds is something done by *human beings*. Hacking therefore rejects Objectivism's uniqueness thesis vis-à-vis natural kinds as senseless, arguing that, "The world is just so complicated that there can be no humanly accessible theory of all of it. No theory can be more than a

9: Philosophical Contexts II

perspective."[30] Just as I have argued throughout this book, Hacking suggests that when we classify, we adopt one among several possible perspectives on the world and in adopting a perspective, we reflect a real portion of that world.

Hacking introduces—as I have done by talking of a synthesis of metaphysics and epistemology—the notion that classification involves a combination of real-world characteristics with human aims and purposes. He believes that natural kinds interest us because of the use we can make of them and the ways in which they can affect us. "We pay attention to passive qualities such as colours and scents because they betoken what we can do with some kinds of things, e.g. eat or escape them," he says.[31] And different people are interested in kinds in different ways. Artisans are interested in those kinds amenable to crafting, farmers are interested in those kinds that can be cultivated or domesticated, and evolutionary biologists are interested in those kinds that exhibit reproductive isolation. Thus, the idea of a single type of natural kind makes no more sense than the idea of a single, all-purpose taxonomy, Hacking claims. "There is no reason to expect that the kinds that matter in the natural history of a region should be identical . . . in extension or logic to the taxa best suited to evolutionary biology."[32]

Hacking—perhaps more than either Kitcher or Dupré—hammers home the extent to which we humans play a vital role in selecting and defining natural kinds. Yet he still retains a strong hold on realism. In this way, his position probably comes closest to my insistence that classification is not simply a matter of reflecting metaphysics, but involves active choices and decisions on the part of the human classifier. But, unlike Hacking, I concentrate purely on scientific classification. To my mind it is most significant not that different types of people are interested in (kinds of) things in different ways, but that even in the much narrower, scientific field, disputes over classification remain rife. Even at the most fundamental level of explanation, it seems that the world is sufficiently complex to admit of differing classifications of the entities within it.

❋

Kitcher, Dupré, Hacking and I share various important beliefs about classification and natural kinds. We all acknowledge the complexity of the natural world. We all agree that this complexity generates plural and competing schemes of classification which reflect different human interests and purposes. And we all insist that several competing classifications can be correct, so long as they reflect patterns existing in the real world. Yet my pluralist account of scientific classification emphasizes aspects which the others do not. I speak about scientists being faced with a *choice* or *decision* when classifying natural entities and I insist that without the necessary choice being made, classification simply cannot get started. I also talk about the synthesis of metaphysics and epistemology, stressing that classification can never be a matter of simply reflecting metaphysics or the way the world is. And, perhaps most importantly, I suggest that the explanation-based view of categorization is an apt model for expert scientific as well as for lay classification. By emphasizing the unifying and motivating role played by underlying theories, aims and explanations, I bring the human factor in classification to the fore in a novel and unique manner.

10

Concluding Remarks: Limitation and Classification

I want to return once again to what, perhaps, will be people's greatest fear—that any pluralist account of scientific classification suggests that scientists describe and dissect a structureless world in any way that takes their fancy. I have already taken considerable pains in previous chapters to insist that scientists' decisions must always be informed by reality—by metaphysics or the way the world is—but it might now be helpful to consider in what specific ways reality constrains classification.

There are (at least) two important ways in which metaphysics binds scientific classification, I suggest—through a limitation of utility and through a 'suitably scientific' limitation. These two limitations are not independent, but are closely linked.

LIMITATIONS ON CLASSIFICATION

Scientists must initially be bound by *what is useful* in the ways in which they classify and divide, meaning that their job is one of uncovering and exploiting regularities which really do occur in nature. It is these regularities which must serve as the basis for any division of the natural world into classes. Scientific classification is, in this way, bound by metaphysics or the way the world is. To discover and exploit regularities in the natural world is to make that world something which we can negotiate or successfully interact with. To know that certain regularities hold between entities is to be able to predict how those entities will behave. To know that a certain substance has atomic number 79, for example, is to know that instances of that substance will behave comparably under certain circumstances. This enables us to foresee that behavior and to act accordingly. On the other hand, to know that two instances of this same substance have different isotope numbers (but the same atomic number) is to know that the two may behave in very different ways under a different set of circumstances or conditions. This again enables us to predict and take account of a different kind of behavior. In each case, some kind of regularity occurs in nature and knowledge of that regularity is useful to us.

The objectivist's mistake, then, does not reside in the claim that classification concerns metaphysics. Of course classification concerns metaphysics—it is the job of science to uncover the regularities and patterns which exist *in reality*. Rather, the objectivist's mistake is to assume that there exists a unique set of regularities and so classes into which the natural world can be divided. *Even at the scientific level*, there exist different patterns and regularities which criss-cross the natural world—and different regularities can result in different divisions of that world. It is not, as the objectivist would have it, that we are untidy in our cognition of the world, but that the world is 'untidy,' its richness eluding the straightjacket of a single, all-embracing system of classification.

Secondly, those properties which scientists deem essential for, or definitive of, a kind must be *suitably scientific*. It is difficult to give a clear definition of 'suitably scientific,' but by this I roughly mean that these properties must reside at a level fundamental enough that they can account for important behavioral characteristics. To group entities together on the basis of prettiness, for example, is unlikely to qualify as a suitably scientific classification. This is so for two reasons. Firstly, entities which are all pretty are unlikely, as a group, to exhibit important regularities or behavioral characteristics beyond their prettiness. In other words, such a grouping would be more-or-less arbitrary from the scientific point of view. What important or significant regularities can we find in the class of a pretty gemstone, a pretty flower and a pretty painting beyond the fact that they are all described as pretty? Secondly, prettiness is a non-measurable characteristic and so is unlikely to qualify as a fundamental property from the scientific point of view. In general, those properties which feature in the scientific classification of entities are, at least to a certain extent, measurable or quantifiable in physical terms in a way that properties such as prettiness are not. It might, therefore, be the case that a property like prettiness can be accounted for by a scientifically more fundamental or significant property, but not vice versa. It may well be that the members of a scientifically relevant class are all judged to be pretty, but this will be explained by the fact that they share some other, more fundamental property or properties.

Different philosophers have used slightly different ways to describe the kind of properties with which scientists concern themselves. D. H. Mellor notes that scientists concentrate on microstructural properties (atomic number for chemical elements and genetic makeup for plants and animals). This is because they use a principle of "microreduction," he explains, which involves the notion that "properties of things should be explained in terms of the properties and relations of their spatial parts."[1] John Dupré defines a natural kind as "a class of objects defined by common possession of some theoretically important property (generally, but not necessarily, microstructural)."[2] Hilary Putnam considers natural kinds to be groups of things which share "important physical properties."[3] Despite the minor differences, these philosophers all agree that a scientific classification is in some way an important and relevant classification, in the sense that those properties deemed to define a particular class will be deeply explanatory of, or fundamental to, that class. All will be suitably scientific properties. In some sense, then, the suitably scientific constraint is a psychological constraint. It involves various people's conceptions[4] of what kinds of property scientists should concern themselves with. It is, however, also a real or physical constraint. Science aims to discover the nature of the world and this nature, it seems, is to be found at a deep, fundamental level.

Of course, none of this means that there exists only one set of scientifically relevant classifications, nor that there exists only one set of suitably scientific properties. As we have seen throughout the course of this book, scientists produce competing classifications using different fundamental properties, and there is no basis on which we can or should proclaim one of these classifications better, more accurate or more scientific than its competitors. The hidden nature of the world proves just as rich and complex as its surface appearance.

Linking the Limitations

How, then, are the two limitations—that scientists concentrate on patterns of regularities which exist in the natural world and that they deal in properties which are suitably scientific—linked? I suggest that suitably scientific properties can be understood as *accounting for* patterns of regularity or behavioral characteristics. We might argue that the reason entities display similar characteristics or behave in similar ways under similar conditions is that they share the same fundamental properties. In other words, the possession of those fundamental properties accounts for those regularities or behavioral characteristics. We might, for example, claim that the reason all samples of gold react in comparable ways when they come into contact with other chemical substances is that all samples of gold have atomic number 79. Possession of fundamental properties determines behavioral characteristics. This, of course, again involves utility—it is useful, in terms of our successful interaction with the world, that we group together entities which react in similar ways, as accounted for by possession of common properties.

Given that suitably scientific properties account for behavior, two important questions arise. Does it make sense to claim that certain behavioral patterns are more basic or fundamental than others? Does it make sense, for example, to claim that those behavioral characteristics associated with atomic number are more basic than those associated with isotope number? And can we capture the *most* basic or fundamental behavioral patterns (if such a set exists) by providing a unique division of entities into kinds on the basis of the most fundamental properties that those entities possess (again, if such a set exists)? The objectivist will want to answer both questions in the affirmative. S/he will say that there *is* one unique division into kinds, based on the most fundamental properties, and that this division captures the most basic behavioral patterns that occur in nature. And this is so, according to the objectivist, regardless of whether or not we can identify the requisite behavioral patterns and properties.

I do not accept this move, however. Even at the scientific level, where classifications are made using important physical properties, or microstructural properties, or theoretically important properties, I suggest that there is not necessarily a unique taxonomy of the natural world. Even at this level, the world still displays an inherent flexibility and indeterminacy. Alternative patterns of behavioral characteristics run through nature and these alternative patterns can be accounted for by different important physical properties, which in turn produce different natural kinds.

10: Concluding Remarks

Keith Donnellan's example using gold, which we discussed in chapter 8, illustrates the point. Recall that Donnellan argued that although we group all isotopes of gold together as one natural kind on the basis of atomic number, it would be equally reasonable and scientific to divide what we call gold into a number of different natural kinds on the basis of different isotope numbers. We can now add that the fact that all isotopes of what we call gold possess atomic number 79, or have 79 protons in the nucleus of the atom, is the important physical property which accounts for their uniform behavior with respect to chemical reactions. But it would also be correct to say that the fact that different isotopes of what we call gold have different combined numbers of protons and neutrons in the nucleus of the atom accounts for their differing behavior regarding radioactivity and chemical breakdown. Both alternatives are reasonable and veracious accounts of the behavioral characteristics of this/these substance(s). Both are scientific, concentrating on microstructural properties to account for important behavioral patterns. Both enable us to better predict and interact with aspects of our environment. How, then, can we say that one is more fundamental, or basic, or scientific an account than the other? We cannot. We must simply admit that the two accounts are different, yet equally legitimate, objective *and correct*. We must also admit that the properties (atomic number and isotope number) that account for these alternative behaviors generate different, yet equally legitimate, objective *and correct* natural kinds.

In one of his recent papers, Hilary Putnam comments that, "Certainly, the Aristotelian insight that objects have structure is right, provided we remember that what counts as the structure of something is relative to the ways in which we interact with it."[5] Of course our scientists do not describe and dissect a structureless world in any way that takes their fancy. They are, after all, interested in reflecting not figments of their imagination, but the way in which the world is. But that world, I submit, is, in reality, so intricately structured that description and dissection of it involves choosing one perspective among several. Which perspective our scientists adopt will dictate both the nature and the boundaries of our natural kinds, but we must remember that this represents a choice between real attributes of real entities residing in a real world. Scientific classification thus involves a synthesis of metaphysics and epistemology. And, as the title of this book proclaims, classification involves the scientist in both discovery *and* decision-making.

Notes

1. Introduction: Complexity and the Natural World

1. See R. N. Giere, *Explaining Science: A Cognitive Approach* (Chicago: University of Chicago Press, 1988), Chapter 1.

2. Objectivism

1. Aristotle, *De Partibus Animalium I*, trans. D. M. Balme (Oxford, England: Clarendon Press, 1963), 642b 21.
2. Ibid., 643a 27–28.
3. Ibid., 634a 32.
4. Ibid., 640a 33–35.
5. Aristotle, *Metaphysics*, 1033b 6–7, in *A New Aristotle Reader*, ed. J. L. Ackrill (Oxford, England: Clarendon Press, 1987).
6. J. Locke, *An Essay Concerning Human Understanding*, ed. A. D. Woozley (Glasgow, Scotland: Collins, 1964), p. 270.
7. L. Wittgenstein, *Tractatus Logico Philosophicus*, trans. D. F. Pears and B. F. McGuinness (London, England: Routledge and Kegan Paul, 1961), paragraphs 2.1 and 2.11.
8. Ibid., paragraph 4.26.
9. The opinions attributed to Putnam in this chapter come from his early strong realist phase. He has since distanced himself from this position, arguing instead for theories such as internal realism, which I discuss in the next chapter.
10. H. Putnam, *Mind, Language, and Reality* (Cambridge, England: Cambridge University Press), p. 233.
11. Ibid., p. 224.
12. He allows for an 'almost all' qualification, which covers the situation where most of the items are gold, but a few are only superficially similar to gold and are not, in fact, gold.
13. S. A. Kripke, Naming and Necessity. In D. Davidson and G. Harman, eds., *Semantics of Natural Language* (Dordrecht, Holland: D. Reidel Publishing Company, 1972), p. 318.

3. Internal Realism I

1. From *Reason, Truth and History* onwards.
2. H. Putnam, *Reason, Truth and History* (Cambridge, England: Cambridge University Press, 1981), p. 49.
3. Ibid., p. 49.
4. It is worth pointing out that a God's Eye perspective will be unattainable even by God. A deity who can stand outside the universe will not see the uniquely correct way in which that universe can be described (since no such description exists), but will see all the numerous ways in which we might correctly describe (portions of) the universe.
5. Putnam, 1981, p. 50.
6. Ibid.
7. See G. Lakoff, *Women, Fire and Dangerous Things: What Categories Reveal about the Mind* (Chicago: University of Chicago Press, 1987), p. 262.
8. Putnam, 1981, p. 52.
9. Ibid., p. 53.
10. Ibid., pp. 49–50.
11. Ibid., p. 73.
12. Ibid., p. 131.
13. Ibid., p. 55.
14. M. Johnson, *The Body in the Mind* (Chicago: University of Chicago Press, 1994), p. 210.
15. Ibid., p. 203.
16. Ibid., p. 211.
17. Ibid., p. 211.

4. Internal Realism II: Criticisms and Implications

1. H. Field, "Realism and Relativism," *Journal of Philosophy* 79(1982): 553–554.
2. Ibid., p. 554.
3. Aristotle, *Metaphysics* 1030a 3 and 1031b 17–18. In *A New Aristotle Reader*, ed. J. L. Ackrill (Oxford, England: Clarendon Press, 1987).
4. Wittgenstein, 1961, paragraph 4.26.
5. Ronald Giere (1988, p. 98) notes that, with respect to the uniqueness requirement, "metaphysical realism plays no role in modern science" and "the rejection of metaphysical realism therefore eliminates nothing that an adequate theory of science might require." As we will see in coming chapters, that is certainly the case with scientific classification. Nonetheless, whatever the situation with science, some philosophers have certainly adhered to metaphysical realism's uniqueness requirement.
6. Putnam, 1981, p. 49.
7. Field, 1982, p. 555.
8. Ibid.
9. Ibid.
10. Field, 1982, pp. 555–556.

11. Putnam, 1981, pp. 72–73.
12. Ibid., p. 556.
13. Ibid., p. 557.
14. Field, 1982, p. 557.
15. See H. Putnam, "Sense, Nonsense and the Senses: An Inquiry into the Powers of the Human Mind," *Journal of Philosophy* 91(1994): 448.
16. H. E. Longino, *Science as Social Knowledge* (Princeton: Princeton University Press, 1990), p. 67.
17. Ibid., p. 73.
18. Ibid., p. 74.
19. Giere, 1988, p. 109.
20. Ibid., p. 6.
21. Ibid., p. 93.
22. Ibid., p. 6.

5. THE PSYCHOLOGY OF CATEGORIZATION

1. See M. E. McCloskey and S. Glucksberg, "National Categories: Well Defined or Fuzzy Sets?" *Memory and Cognition* 6(1978): 462–472.
2. See, for instance, E. Rosch and C. B. Mervis, "Family Resemblances: Studies in the Internal Structure of Categories," *Cognitive Psychology* 7(1975): 573–605; E. Rosch, "On the Internal Structure of Perceptual and Semantic Categories." In T. E. Moore, ed., *Cognitive Development and the Acquisition of Language* (San Diego, CA: Academic Press, 1973), 111–145.
3. See F. S. Bellezza, "Reliability of Retrieval from Semantic Memory: Common Categories," *Bulletin of the Psychonomic Society* 22(1984): 324–326.
4. See, for instance, Rosch and Mervis, 1975.
5. See Rosch and Mervis, 1975.
6. At roughly this time, other psychologists propounded the exemplar view, which holds that concepts are mentally represented as specific instances or exemplars, rather than as abstract summary representations. I am excluding discussion of this view because it simply constitutes another reaction against the classical view, with similar aims to the probabilistic view.
7. See G. L. Murphy and D. L. Medin, "The Role of Theories in Conceptual Structure," *Psychological Review* 92(1985): 289–314, for one of the earliest expressions of this doubt.
8. See N. Goodman, *Problems and Projects* (Indianapolis, IN: Bobbs-Merrill, 1979), 24–32, for parallel views about similarity from a philosopher.
9. Example taken from L. K. Komatsu, "Recent Views of Conceptual Structure," *Psychological Bulletin* 112(1992): 505.
10. See L. J. Rips, Similarity, Typicality and Categorization. In S. Vosinadou and A. Ortony, eds., *Similarity and Analogical Reasoning* (Cambridge, England: Cambridge University Press, 1989), 21–59.
11. Rips is not talking here about metaphysically essential and accidental properties. Since psychology deals with epistemology and not metaphysics, Rips is simply aiming to distinguish changes which *we might consider fundamental* from those which

we might not consider fundamental in relation to categorization judgements, irrespective of whether or not there are such things as metaphysically essential properties.

12. Keil conducted a similar experiment in which subjects were told about a raccoon which underwent surgery in order to make it look and behave as a skunk would. The subjects judged that, despite these superficial changes, the animal was still a raccoon and not a skunk. (F. C. Keil, *Concepts, Kinds and Cognitive Development* (Cambridge: MIT Press, 1989).)

13. W. D. Wattenmaker, G. V. Nakamura, and D. L. Medin, Relationships between Similarity-based and Explanation-based Categorization. In D. J. Hilton, ed., *Contemporary Science and Natural Explanation* (Brighton, England: Harvester, 1988), 204–240.

14. Various pieces of empirical evidence go toward illustrating that our concepts are underlain and driven by our knowledge and theories about the world, but it is beyond the scope of this chapter to discuss them. See Komatsu (1992) and Medin (1989) for summaries of some of the evidence.

15. D. L. Medin and A. Ortony, Psychological Essentialism, in S. Vosinadou and A. Ortony, eds., *Similarity and Analogical Reasoning* (Cambridge, England: Cambridge University Press, 1989), 179–195.

16. There are, in fact, various models which exhibit an explanation-based structure. See Komatsu (1992) for a summary.

17. McNamara and Sternberg (1983) provide experimental evidence showing that people tend to believe in necessary and sufficient defining conditions, even though they are not always able to say what these conditions are.

18. Medin and Ortony, 1989, p. 183.

19. The same point can be made regarding Keil's skunk/raccoon experiment. And, more recently, Malt (1990) reports an experiment in which subjects were informed of a plant halfway between a marigold and a dandelion in appearance. They judged that it would make more sense to say, "We'd have to ask an expert to tell us which it is" than to say, "I guess you can call it whichever you want." This suggests that the subjects believed the plant to possess some hidden essence in virtue of which it would fall into one category or the other, despite its apparent ambiguity.

20. See Murphy and Medin, 1985.

21. D. L. Medin, "Concepts and Conceptual Structure," *American Psychologist*, 44(1989): 1469–1481.

6. PHILOSOPHY AND THE PSYCHOLOGY OF CATEGORIZATION

1. This theory was not discussed in chapter 5. This is because (a) as an attempt to combine the classical and probabilistic views, it adds nothing to a general overview of the history of the psychology of categorization and (b) this theory won little support from the psychological community as a whole.

2. E. E. Smith, D. L. Medin and L. J. Rips, "A Psychological Approach to Concepts: Comments on Rey's 'Concepts and Stereotypes,'" *Cognition* 17(1984): 267–268.

3. G. Rey, "Concepts and Stereotypes," *Cognition* 15(1983): 243.

4. Ibid.

5. Ibid., p. 251.

6. Ibid., p. 255.

7. Many experiments have required subjects to list attributes they believe to be characteristic of a class of objects. At least some of the attributes listed are mentioned by the majority of subjects in each experiment and these common attributes also appear across different experiments. Smith, Medin and Rips cite Malt and Smith (1982), Rosch and Mervis (1975) and Smith et al. (1984) in support of their claim.

8. Medin, 1989, p. 1469.

7. Five Interrelated Theses: The Theory Explained

1. See chapter 8 for discussion of the contemporary dispute.

2. A. J. Desmond, "Designing the Dinosaur: Richard Owen's Response to Robert Edmond Grant," *Isis* 70(1979): 229.

3. Desmond informs us that Owen was incensed by Grant's attempt to construct an historical Lamarckism and that the two men came into conflict publicly in the 1830s. Owen yielded much power in both science and society and, in the mid 1830s, he prevented Grant's appointment as comparative anatomist at the Zoological Society of London. When dealing with the Lamarckian problem, Owen quoted from some of Grant's lectures, considering him the latest upholder of transmutation in relation to paleontology.

4. Owen actually tested Grant's historical Lamarckian hypothesis, but found no support for it. The New Red Sandstone rocks should have been home to the most primitive and ancient amphibians, yet the fossil forms found there were clearly more advanced than currently existing salamanders, frogs and apodans. It seems that, contrary to Grant's Lamarckian hypothesis, these forms first made their appearance in their highest, rather than in their lowest, developmental state. Furthermore, fossil evidence showed that lizards were more ancient than the amphibians which were supposed to be their predecessors, according to the Lamarckian scale.

5. As we will see in chapter 8, common ancestry forms the basis of the contemporary cladistic or phylogenetic species concept.

6. M. P. Winsor, "Barnacle Larvae in the Nineteenth Century: A Case Study in Taxonomic Theory," *Journal of the History of Medicine* 24(1969): 309.

7. At this time, Thompson was unaware of the fact that barnacles first hatch out as what are now called nauplii and then later change into a different form now known as the cypris larva. So, in the 1826 case, Thompson had *actually* seen cypris larva change into sessile barnacles, while in 1830, he saw nauplii hatch from the eggs of stalked barnacles. Since he was not aware of the two stages in the life of a young barnacle, he attributed the difference to some major discrepancy between the two forms of barnacle.

8. Winsor, 1969, p. 299.

9. Prior to this revision, he had stated that in the future it might be necessary to classify the Cirripedia with the Crustacea, but that currently the scientific data was not adequate to make such a decision.

10. Winsor, 1969, p. 301.

8. Philosophical Contexts I: Critiques of Natural Kinds

1. Putnam, 1975, pp. 238–39.
2. Donnellan points out that, in actuality, isotopes are distinguished by a combination of atomic and isotope number, but for the sake of simplicity, he assumes that they are distinguished by isotope number alone. Either way, the points he goes on to make remain the same.
3. Donnellan indicates that he does not know whether this is actually the case on Earth but, while this is not at all necessary for the story to carry through, it does increase its psychological plausibility.
4. K. S. Donnellan, Kripe and Putnam on Natural Kind Terms. In C. Ginet and S. Shoemaker, eds., *Knowledge and Mind: Philosophical Essays* (Oxford, England: Oxford University Press, 1983), p. 104.
5. Canfield uses the term 'intentional' throughout—as opposed to the usual 'intensional'—in contradistinction with 'extensional.' His use of 'intentional' conveys the notion that intentional discoveries are discoveries that we make *knowingly*. Thus Canfield explains that although Columbus actually discovered a new land, we would not say he discovered *that* the place he sailed to was a new land since he believed he had sailed to the Andes.
6. J. V. Canfield, Discovering Essence, in C. Ginet and S. Shoemaker, eds., *Knowledge and Mind: Philosophical Essays* (Oxford, England: Oxford University Press, 1983), p. 107.
7. See chapter 2 for some examples of this kind of strategy.
8. Canfield, 1983, p. 118.
9. They also back up their accounts with carefully chosen examples—such as water and gold—in which (a) the definition is well-established, and (b) there is just one extant definition and so no question of multiple or conflicting candidates.
10. Canfield, 1983, p. 119.
11. S. S. Hughes, *The Virus: A History of the Concept* (London, England: Heinemann Educational, 1977), p. 113.
12. Ibid., pp. 113–14.
13. Canfield, 1983, p. 113.

9. Philosophical Contexts II: Contemporary Philosophy of Biology

1. For clarity and ease of understanding, I paint the schools themselves and their differences in very broad brush strokes. See Hull (1988) for an in-depth look at the development of, diversity within and differences between these schools.
2. Hull, 1988, p. 119.
3. J. Dupré, *The Disorder of Things* (Cambridge: Harvard University Press, 1993), p. 45.
4. Mayr, 1963, p. 19 (quoted in Sokal and Crovello, 1992).

Notes

5. E. Mayr, Species Concepts and their Application. In E. Sober, ed., *Conceptual Issues in Evolutionary Biology* (Cambridge: MIT Press, 1984), p. 533.

6. Ibid., p. 531.

7. Mishler and Donoghue (1992) list similar problem cases for the biological species concept. They also note that while proponents of the concept consider that gene flow maintains the cohesion of the species, there are other factors which may also confer cohesion, such as internal homeostasis or the state of the external environment. They therefore conclude that it is not possible to give a univocal statement concerning cohesion and its causes.

8. Mayr, 1984, p. 537.

9. See R. R. Sokal and T. J. Crovello, The Biological Species Concept: A Critical Evaluation. In M. Ereshefsky, ed., *The Units of Evolution* (Cambridge: MIT Press, 1992), 27–55.

10. Ibid., p. 46.

11. See, for example, Hull, 1988, chapter 7 and Hull, 1989, chapter 10.

12. K. De Queiroz, "Phylogenetic Definitions and Taxonomic Philosophy," *Biology and Philosophy* 7(1992): 300.

13. G. G. Simpson, by contrast, argues that speciation occurs when a descendant population becomes reproductively isolated from its ancestors and not when an ancestral population gives rise to two descendant populations which are reproductively isolated from one another. (See P. Kitcher, "Species," *Philosophy of Science* 51(1984a): 308–333.)

14. Borrowed from E. Sober, *Philosophy of Biology* (Oxford, England: Oxford University Press, 1993).

15. Dupré, 1993, p. 48.

16. Nelson and Platnick, 1981, p. 12 (quoted in Hull, 1988, p. 248).

17. Ibid., p. 35 (quoted in Hull, 1988, p. 248).

18. Ibid., p. 139 (quoted in Hull, 1989, p. 152).

19. Ibid., p. 14 (quoted in Hull, 1989, p. 152).

20. Further divergent views exist, such as that of van Valen, which takes speciation to be the process by which descendant populations are ecologically differentiated from their ancestors. (See Kitcher, 1984a.)

21. See Craraft, 1983.

22. Sokal and Sneath, 1963, p. 265 (quoted in Hull, 1988, chapter 4).

23. See Kitcher 1984a and 1984b; Dupré 1981 and 1993.

24. Kitcher, 1984a, p. 330.

25. Ibid., p. 319.

26. Kitcher, 1984b, p. 630.

27. Dupré, 1993, p. 57.

28. Ibid., p. 53.

29. Ibid., p. 58.

30. I. Hacking, Natural Kinds. In R. B. Barrett and R. F. Gibson, eds., *Perspectives on Quine* (Oxford, England: Basil Blackwell, 1990), p. 139.

31. I. Hacking, "A Tradition of Natural Kinds," *Philosophical Studies* 62(1991): 113–14.

32. Ibid., p. 123.

10. Concluding Remarks: Limitation and Classification

1. D. H. Mellor, "Natural Kinds," *British Journal for the Philosophy of Science* 28 (1977): 310.
2. J. Dupré, "Natural Kinds and Biological Taxa," *Philosophical Review* 90 (1981): 68.
3. See, for instance, Putnam, 1975, p. 239.
4. Those of scientists, philosophers and perhaps even laypeople.
5. Putnam, 1993, p. 134.

Bibliography

Ackrill, J. L., ed. *A New Aristotle Reader.* Oxford, England: Clarendon Press, 1987.

Aristotle. *Categories and De Interpretatione.* Translated by J. L. Ackrill. Oxford, England: Clarendon Press, 1963.

Aristotle. *De Partibus Animalium I and De Generatione Animalium I.* Translated by D. M. Balme. Oxford, England: Clarendon Press, 1972.

Barsalou, L. W. Intraconcept Similarity and its Implications for Interconcept Similarity. In S. Vosniadou and A. Ortony, eds. *Similarity and Analogical Reasoning.* Cambridge, England: Cambridge University Press, 1989: 76–121.

Bellezza, F. S. "Reliability of Retrieval from Semantic Memory: Noun Meanings." *Bulletin of the Psychonomic Society* 22(1984): 377–380.

Bellezza, F. S. "Reliability of Retrieval from Semantic Memory: Common Categories." *Bulletin of the Psychonomic Society* 22(1984): 324–326.

Canfield, J. V. Discovering Essence. In C. Ginet and S. Shoemaker, eds. *Knowledge and Mind: Philosophical Essays.* Oxford, England: Oxford University Press, 1983: 105–129.

Churchland, P. S. *Neurophilosophy.* Cambridge: MIT Press, 1986.

Cracraft, J. "Species Concepts and Speciation Analysis." *Current Ornithology* 1(1983): 159–187.

De Queiroz, K. "Phylogenetic Definitions and Taxonomic Philosophy." *Biology and Philosophy* 7(1992): 295–313.

Dean, J. Controversy Over Classification: A Case Study from the History of Botany. In B. Barnes and S. Shapin, eds. *Natural Order.* London, England: Sage Publications, 1979: 211–230.

Desmond, A. J. "Designing the Dinosaur: Richard Owen's Response to Robert Edmond Grant." *Isis* 70(1979): 224–234.

Donnellan, K. S. Kripke and Putnam on Natural Kind Terms. In C. Ginet and S. Shoemaker, eds. *Knowledge and Mind: Philosophical Essays.* Oxford, England: Oxford University Press, 1983: 84–104.

Dupré, J. "Natural Kinds and Biological Taxa." *Philosophical Review* 90(1981): 66–90.

Dupré, J. *The Disorder of Things.* Cambridge: Harvard University Press, 1993.

Eddings, D. *The Elenium.* London, England: HarperCollins, 1993.

Ereshefsky, M. *The Units of Evolution.* Cambridge: MIT Press, 1992.

Field, H. "Realism and Relativism." *Journal of Philosophy* 79(1982): 553–567.

Giere, R. N. *Explaining Science: A Cognitive Approach.* Chicago: University of Chicago Press, 1988.

Goodman, N. Seven Strictures on Similarity. In N. Goodman. *Problems and Projects*. Indianapolis, IN: Bobbs-Merrill, 1979: 24–32.

Hacking, I. Natural Kinds. In R. B. Barrett and R. F. Gibson, eds. *Perspectives on Quine*. Oxford, England: Basil Blackwell, 1990: 129–141.

Hacking, I. "A Tradition of Natural Kinds." *Philosophical Studies* 61(1991): 109–26.

Hughes, S. S. *The Virus: A History of the Concept*. London, England: Heinemann Educational, 1977.

Hull, D. L. A Matter of Individuality. In E. Sober, ed. *Conceptual Issues in Evolutionary Biology*. Cambridge: MIT Press, 1984: 623–645.

Hull, D. L. *Science as a Process*. Chicago: University of Chicago Press, 1988.

Hull, D. L. *The Metaphysics of Evolution*. Albany: SUNY Press, 1989.

Johnson, M. *The Body in the Mind*. Chicago: University of Chicago Press, 1994.

Keil, F. C. *Concepts, Kinds and Cognitive Development*. Cambridge: MIT Press, 1989.

Kitcher, P. "Species." *Philosophy of Science* 51(1984a): 308–333.

Kitcher, P. "Against the Monism of the Moment: A Reply to Elliott Sober." *Philosophy of Science* 51(1984b): 616–630.

Komatsu, L. K. "Recent Views of Conceptual Structure." *Psychological Bulletin* 112 (1992): 500–526.

Kripke, S. A. Naming and Necessity. In D. Davidson and G. Harman, eds. *Semantics of Natural Language*. Dordrecht, Holland: D. Reidel Publishing Company, 1972: 253–355.

Lakoff, G. *Women, Fire and Dangerous Things: What Categories Reveal about the Mind*. Chicago: University of Chicago Press, 1987.

Lakoff, G. and Johnson, M. *Metaphors We Live By*. Chicago: University of Chicago Press, 1980.

Locke, J. *An Essay Concerning Human Understanding*. Edited by A. D. Woozley. Glasgow, Scotland: Collins, 1964.

Longino, H. E. *Science as Social Knowledge*. Princeton: Princeton University Press, 1990.

Malt, B. C. "Features and Beliefs in the Mental Representation of Categories." *Journal of Memory and Language* 29(1990): 289–315.

Malt, B. C. "Water is not H_2O." *Cognitive Psychology* 27(1994): 41–70.

Malt, B. C. and Smith, E. E. "The Role of Familiarity in Determining Typicality." *Memory and Cognition* 10(1982): 69–75.

Mayr, E. Species Concepts and their Application. In E. Sober, ed. *Conceptual Issues in Evolutionary Biology*. Cambridge: MIT Press, 1984: 531–540.

McCloskey, M. E. and Glucksberg, S. "Natural Categories: Well Defined or Fuzzy Sets?" *Memory and Cognition* 6(1978): 462–472.

McNamara, T. P. and Sternberg, R. J. "Mental Models of Word Meaning." *Journal of Verbal Learning and Verbal Behaviour* 22(1983): 449–474.

Medin, D. L. "Concepts and Conceptual Structure." *American Psychologist* 44(1989): 1469–1481.

Medin, D. and Ortony, A. Psychological Essentialism. In S. Vosniadou and A. Ortony, eds. *Similarity and Analogical Reasoning*. Cambridge, England: Cambridge University Press, 1989: 179–195.

Mellor, D. H. "Natural Kinds." *British Journal for the Philosophy of Science* 28(1977): 299–312.

Mervis, C. B. and Pani, J. R. "Acquisition of Basic Object Categories." *Cognitive Psychology* 12(1980): 496–522.

Mishler, B. D. and Donoghue, M. J. Species Concepts: A Case for Pluralism. In M. Ereshefsky, ed. *The Units of Evolution*. Cambridge: MIT Press, 1992: 121–137.

Murphy, G. L. and Medin, D. L. "The Role of Theories in Conceptual Structure." *Psychological Review* 92(1985): 289–314.

Putnam, H. *Mind, Language and Reality*. Cambridge, England: Cambridge University Press, 1975.

Putnam, H. *Reason, Truth and History*. Cambridge, England: Cambridge University Press, 1981.

Putnam, H. *Realism with a Human Face*. Cambridge: Harvard University Press, 1990.

Putnam, H. Aristotle after Wittgenstein. In R. W. Sharples, ed. *Modern Thinkers and Ancient Thinkers*. London, England: UCL Press, 1993: 117–137.

Putnam, H. "Sense, Nonsense and the Senses: an Inquiry into the Powers of the Human Mind." *Journal of Philosophy* 91(1994): 445–517.

Rey, G. "Concepts and Stereotypes." *Cognition* 15(1983): 237–262.

Rey, G. "Concepts and Conceptions: A Reply to Smith, Medin and Rips." *Cognition* 19(1985): 297–303.

Rips, L. J. Similarity, Typicality and Categorisation. In S. Vosniadou and A. Ortony, eds. *Similarity and Analogical Reasoning*. Cambridge, England: Cambridge University Press, 1989: 21–59.

Rosch, E. On the Internal Structure of Perceptual and Semantic Categories. In T. E. Moore, ed. *Cognitive Development and the Acquisition of Language*. San Diego, CA: Academic Press, 1973: 111–145.

Rosch, E. Principles of Categorization. In E. Rosch and B. Lloyd, eds. *Cognition and Categorization*. Hillsdale, NJ: Lawrence Erlbaum, 1978: 27–48.

Rosch, E. and Mervis, C. B. "Family Resemblances: Studies in the Internal Structure of Categories." *Cognitive Psychology* 7(1975): 573–605.

Smith, E. E. "Three Distinctions about Concepts and Categorization." *Mind and Language* 4(1989): 57–61.

Smith, E. E. and Medin, D. L. *Categories and Concepts*. Cambridge: Harvard University Press, 1981.

Smith, E. E., Medin, D. L. and Rips, L. J. "A Psychological Approach to Concepts: Comments on Rey's 'Concepts and Stereotypes.'" *Cognition* 17(1984): 265–274.

Smith, E. E., Osherson, D. N., Rips, L. J. and Keane, M. *A Theory of Conceptual Combination for Prototype Concepts*. Unpublished manuscript, Massachusetts Institute of Technology, 1984.

Sober, E. *Philosophy of Biology*. Oxford, England: Oxford University Press, 1993.

Sokal, R. R. and Crovello, T. J. The Biological Species Concept: A Critical Evaluation. In M. Ereshefsky, ed. *The Units of Evolution*. Cambridge: MIT Press, 1992: 27–55.

Staniland, H. *Universals*. Garden City, NY: Doubleday, 1972.

Turbocharged Dinosaur. (21 January 1999). On world wide web: http://news.bbc.co.uk/hi/english/sci/tech/newsid_259000/259902.stm.

Wattenmaker, W. D., Nakamura, G. V. and Medin, D. L. Relationships between Similarity-based and Explanation-based Categorization. In D. J. Hilton, ed. *Contemporary Science and Natural Explanation*. Brighton, England: Harvester, 1988: 204–240.

Whitehouse, D. (21 January 1999). Pluto Will Have 'Dual Citizenship.' On world wide web: http://news.bbc.co.uk/hi/english/sci/tech/newsid_259000/259767.stm.

Winsor, M. P. "Barnacle Larvae in the Nineteenth Century: A Case Study in Taxonomic Theory." *Journal of the History of Medicine* 24(1969): 294–309.

Wittgenstein, L. *Tractatus Logico Philosophicus*. Translated by D. F. Pears and B. F. McGuinness. London, England: Routledge and Kegan Paul, 1961.

Index

Ampère, 84
anagenetic speciation, 104
Aristotle, 23–24, 39

Bellezza, Francis, 53–54
biological species concept, 100–102; problems with, 100–102, 106

Canfield, John, 92–97
case studies (used to illustrate scientific classification), 75–87
categories: existing objectively, 21; logical relations between, 21
categorization: classical view. See classical view (of categorization); explanation-based account. See explanation-based account (of categorization); family resemblance view. See probabilistic view (of categorization); human contribution to, 65, 73–87, 91, 96; judgments differ from similarity judgments, 57–58; probabilistic view. See probabilistic view (of categorization); psychology of, 18, 51–65; psychology of, and philosophy, 66–72; psychology of, and Rey's attack, 67–70
choice: by scientists, 17, 29–30, 45–46, 49–50, 73–87, 90–91, 95, 96, 107, 110, 115–16
cladism. See phylogeny
classical view (of categorization), 51–54, 64–65; and conceptual categories, 22; and natural kinds, 21; and necessary and sufficient conditions, 51–52, 53–54; and similarity, 56; problems with, 52–54
classification: pluralist account of, 17, 73–87; scientific, comprises synthesis of metaphysics and epistemology, 17, 29–30, 45, 73–74, 75, 79–78, 80, 87, 93, 96–97, 110; scientific, five interrelated theses, 73–74; scientific, limitations on, 111–13
communality (of concepts), 71
concepts: epistemological role of, 68; metaphysical role of, 68; not only represented by feature lists, 58–59, 62; objectivist, 22; and relational knowledge, 59
continuum (of accessibility), 63
core and identification procedure, 66–67, 68
Cracraft, Joel, 104
Crovello, Theodore, 101

Darwin, Charles, 85–87
Dean, John (case study of botanical classification), 75–78
decision, by scientists. See choice, by scientists
degeneration (of reptilian form), 79–81
de Queiroz, Kevin, 102
Desmond, Adrian J. (case study of dinosaur classification), 78–82
dinosaur category, "creation" of, 44, 78–82
Discontinuous Progression, 81
Donnellan, Keith, 88–92, 115; critique of natural kinds, 88–92
Dupré, John, 103, 107–8, 113; promiscuous realism, 107–8

empirical evidence, use of, 46–47
epistemology: objectivist, 22–23; objectivist, failings of, 29–37;

epistemology (*continued*): objectivist, independence from human mind, 23
essence: criterial theory of (Canfield), 92–97; discovery of, 95–96; human belief that objects possess, 62–63; placeholder, 63–64; presupposition of theory of, 93, 94–95
essentialism: doctrine of, 16–17; and natural kinds, 21; psychological, 62–64; and science, 16
explanation-based account (of categorization), 59–64, 65; as model of expert scientific classification, 17, 19, 50, 73–87, 110; psychological essentialism. *See* essentialism, psychological; relational account, 60–61; and similarity, 61–62; and theories of the world, 61–62
extensional discovery, 92–93
External Definitions, Hypothesis of (Rey), 69

family resemblance view. *See* probabilistic view (of categorization)
Field, Hartry, 38–45

Giere, Ronald: constructive realism, 49–50
Glucksberg, Sam, 52
Grant, Robert Edmond, 79–82

Hacking, Ian: theory of natural kinds, 108–9
Hennig, Willi, 102–3
heuristic, underlying categorization, 37, 64, 74

intentional discovery, 92–93
internal realism, 17, 29–37; criticisms of, 38–45; and pluralist account of scientific classification, 45; and reality, 31

Johnson, Mark, 35; theory of truth, 35–37

Kitcher, Philip: pluralistic realism, 105–7

Kripke, Saul, 16, 39–40; Canfield's critique of, 92–97; and objectivist interpretation of science, 25–27

Lamarck, 82
Lamarckism (transmutation), 79–80
language, objectivist account of, 23
limitation (on scientific classification), 111–13; suitably scientific, 112–13; of utility, 111–12
Linnaeus, 82
Locke, John, 24, 90
Longino, Helen: contextualist model of science, 47–48

Martin-Saint-Ange, 84–85
Mayr, Ernst, 100
McCloskey, Michael, 52
meaning: and internal realism, 31–32; objectivist account of, 23
Medin, Douglas, 56–57, 61, 62, 64, 65, 70
Mellor, D. H., 113
Mervis, Carolyn, 55
metaphysical realism, 30, 38, 42; and one true description of the world, 39–41
metaphysics: objectivist, 20–22
methodology, used in this book, 18–19
Milne-Edwards, Henri, 83
monophyletic taxa, 103
Murphy, Gregory, 56–57, 64

Nakamura, Glenn, 61
natural world: complexity of, 16–17, 74, 87, 92, 107–8
Nelson, Gareth, 104
numerical taxonomy. *See* phenetics

Objectivism: doctrine of, 20–28; and essential and accidental properties, 20; and necessary and sufficient conditions, 20–21; and one true description of the world, 39–41; proponents of, 23–27, 66, 69–70, 71; and real-world atomism, 21
objects: "making" of, 43–45

Index

Ortony, Andrew, 62
Owen, Richard, 44, 78–82, 85

pattern cladism, 104
perspective, adoption of by scientists, 45–46
phenetics, 99; and similarity, 99
philosophy, and the psychology of categorization, 66–72
phylogeny, 102–5; advantages and disadvantages, 104–5
Platnick, Nelson, 104
pluralist account (of classification). *See* classification, pluralist account of
pluralistic realism (Kitcher), 105–7
probabilistic view (of categorization), 54–56, 65; and similarity, 56
promiscuous realism (Dupré), 107–8
properties: core, 66–67, 71, 72; identification, 66–67, 72
Putnam, Hilary, 16, 39–40, 113, 115; Canfield's critique of, 92–97; Donnellan's critique of, 88–92; and objectivist interpretation of science, 25–26; response to criticism of internal realism, 40–41

real world, independence from human mind, 20
reflection (of structure of natural world), 20, 22, 74
relations: between concepts, 59, 60; between features (of a concept), 59
Rey, Georges: attack on psychology of categorization, 67–70; Hypothesis of External Definitions, 69

Rips, Lance, 57, 67
Rosch, Eleanor, 52, 55

similarity: in classical and probabilistic views, 56; does not fully explain categorization, 56–59; in explanation-based account, 61–62; judgments differ from categorization judgments, 57–58; overall (phenetics), 99; in a particular context, 56–57
sister-group relationship, 103
Smith, Edward, 66, 70
Sneath, Peter, 99
Sokal, Robert, 99, 101

taxonomy: biological species concept. *See* biological species concept; experimental (biosystematics), 76–77; numerical. *See* phenetics; orthodox (Linnean), 75–76; phenetics. *See* phenetics; phylogeny. *See* phylogeny
theories (explanation-based account), 61–62
Thompson, John Vaughan, 82–87
truth: correspondence theory of, problem with, 42–43; as idealized rational acceptability, 32–33; and internal realism, 30–31, 32–37; more than one version of, 31, 94, 108, 110
Twin Earth (Putnam and Donnellan), 88–92

Wattenmaker, William, 61
Winsor, Mary P. (case study of barnacle classification), 82–87
Wittgenstein, Ludwig, 24–25, 40, 54

```
BD 241 .B785 2000
Bryant, Rebecca, 1970-
Discovery and decision
```